THE SECRET LIFE OF DATA

INFORMATION SOCIETY SERIES

Laura DeNardis and Michael Zimmer

THE SECRET LIFE OF DATA

Navigating Hype and Uncertainty in the Age of
Algorithmic Surveillance

ARAM SINNREICH AND JESSE GILBERT

The MIT Press
Cambridge, Massachusetts
London, England

The MIT Press would like to thank the anonymous peer reviewers who provided comments on drafts of this book. The generous work of academic experts is essential for establishing the authority and quality of our publications. We acknowledge with gratitude the contributions of these otherwise uncredited readers.

This book was set in Adobe Garamond Pro by New Best-set Typesetters Ltd. Printed and bound in the United States of America.

Library of Congress Cataloging-in-Publication Data

Names: Sinnreich, Aram, author. | Gilbert, Jesse, author.
Title: The secret life of data : navigating hype and uncertainty in the age of algorithmic
    surveillance / Aram Sinnreich and Jesse Gilbert.
Description: Cambridge, Massachusetts : The MIT Press, [2024]. | Series: The information
    society series | Includes bibliographical references and index.
Identifiers: LCCN 2023023801 (print) | LCCN 2023023802 (ebook) |
    ISBN 9780262048811 (hardcover) | ISBN 9780262377812 (epub) |
    ISBN 9780262377805 (pdf)
Subjects: LCSH: Data privacy. | Metadata—Social aspects. | Big data—Social aspects. |
    Information society.
Classification: LCC HD30.3815 .S566 2024 (print) | LCC HD30.3815 (ebook) |
    DDC 323.44/8—dc23/eng/20231219
LC record available at https://lccn.loc.gov/2023023801
LC ebook record available at https://lccn.loc.gov/2023023802

10 9 8 7 6 5 4 3 2 1

One can never second guess themselves
too much; there is no such thing
as being right; and answers,
if such a thing ever existed,
are overrated; you are never
where you should be,
only where you are;
faith is for people
who already have it.
—Norman Savage, *A Sunday Sermon for the Broke & Broken; for Those at War*

Contents

Introduction

In 2018, a violent criminal who had terrorized California residents for over a decade was finally brought to justice, more than thirty years after his last known crime. Joseph James DeAngelo, a former police officer, was arrested by Sacramento County sheriff's deputies and charged with a dozen murders attributed to an elusive figure previously known only as the Golden State Killer. After troves of genetic evidence were presented at his trial, DeAngelo pleaded guilty to thirteen counts of first-degree murder and was sent to prison for the remainder of his natural life.

According to news coverage at the time of DeAngelo's arrest, the long-cold case had finally been cracked thanks to technological advances. Old DNA samples were analyzed using modern genetic profiling techniques and the results uploaded to an open-source genealogy website called GEDmatch, where investigators homed in on the killer's identity by tracing his family tree through profiles posted by other users on the site.

Data was hailed as the hero in this case; without the invisible strands of ones and zeroes strung between our bodies, our identities, our markets, and our law enforcement system, justice would never have been served and the Golden State Killer would have remained at large. Even if genealogy websites weren't built for the purpose of aiding in criminal justice, outcomes like this have been trumpeted as examples of how new methods of data collection, storage, and analysis could make the world a better place by bringing clarity and accountability to antiquated systems hindered by guesswork, bias, and obscurity.

"It is fitting that today is National DNA Day," District Attorney Anne Marie Schubert said at a press conference announcing the arrest on April 25, 2018. "We found the needle in the haystack."[1]

Yet, while many rejoiced at DeAngelo's capture and conviction, not everyone found comfort in the investigators' methods. As soon as the news broke, technology ethicists and civil rights advocates began to raise concerns about the broader implications of using ancestry databases to track suspected criminals. Should law enforcement have free rein to use DNA matching techniques in all cases, even those involving nonviolent crimes? Without adequate regulatory and judicial oversight, could techniques like this be used in a sloppy or even malicious fashion, potentially implicating blameless suspects? Should users of open-source databases like GEDmatch be informed that their contributions might unwittingly make them "genetic informants on their innocent family," as attorney Steve Mercer of the Maryland public defender's office framed it in a *USA Today* interview?[2]

The legal and ethical waters became even muddier in 2020, when a *Los Angeles Times* investigation revealed that the official story about DeAngelo's identification and capture had omitted some important details.[3] For one thing, law enforcement investigators didn't collect data only from the open-source GEDmatch site. They also obtained data using a fake account on the private genealogy service FamilyTreeDNA, a potential violation of the company's terms of service. Additionally, police collaborated with a "civilian geneticist" who used her personal account on another private service, MyHeritage, to upload DeAngelo's DNA and explore his family tree without notifying the company of her true intentions. None of this data collection and analysis was undertaken with a warrant, which means that, contrary to the pledges they'd made to their customers, FamilyTreeDNA and MyHeritage shared personal genetic information with police even when they weren't required to by law.

Despite these revelations, many people still believed the benefits of this foray into an ethical gray area were worth the risks. As the daughter of one of DeAngelo's victims told the *Los Angeles Times*, she didn't mind that "practices were bent," because the unmasking of the killer justified any potential violation of privacy, and "the truth is what's important."[4]

Less than two years later, another breaking news story raised the stakes even further, leading many people to reevaluate the relative merits of "truth" and privacy. In 2022, the US Supreme Court issued its landmark decision in *Dobbs v. Jackson*, effectively ending half a century of safe and legal abortion throughout the United States and opening the door for numerous state laws that not only outlawed abortions but treated women who sought them as murderers.[5] Privacy advocates and supporters of abortion rights immediately began to raise the alarm about the risks of online data exposing abortion seekers to liability under these new laws.[6] Simultaneously, many news outlets and nonprofit organizations began public awareness campaigns aimed at showing abortion seekers how to protect their "digital footprints" and thereby minimize their potential exposure to law enforcement.

These concerns were not mere speculation or hyperbole. Even prior to the *Dobbs* decision, there were dozens of cases in the United States in which people's online search histories, chat archives, and other digital traces were used to prosecute them for terminating pregnancies—and even more such cases elsewhere around the world, especially in countries where abortion is not legal. In the wake of *Dobbs*, civil rights watchdogs began to look more closely at a newer set of risks for digital exposure—women's health-tracking apps and biometric devices, such as menstrual calendars and smartwatches. The same constellation of biotechnology, digital databases, identity-tracking business models, and evolving norms about data privacy that had led to the Golden State Killer's arrest now threatened to expose millions of American women to prosecution for seeking a form of medical intervention they'd understood as a fundamental right throughout their entire lives.

The tensions inherent in these two recent news events—between public and private, physical and digital, legal and ethical—are at the heart of this book. Together, they exemplify the double-edged sword of life in the twenty-first century: a potent mixture of possibilities and perils emerging in a data-rich, globally connected society. Armed with smartwatches, home internet appliances, and an endless supply of apps, games, and streaming services, billions of us now spend our days and nights enmeshed in webs of digital sensors, machine learning algorithms, and overlapping information networks, all designed to reduce the minutiae of our lives into discrete data points. These data are then transmitted, stored, and analyzed for the purposes of predicting, evaluating, and influencing our behaviors in a never-ending feedback loop.

While these new data systems have likely improved our lives in countless ways, they have also presented us with a host of new conflicts and conundrums that we still lack the ability—or even the conceptual language—to resolve. How can a MyHeritage user be expected to anticipate the risk that they may become a "genetic informant" on a distant, unknown relative simply by submitting their own DNA to explore their family tree? How can a judge issue a warrant allowing police to investigate an individual suspect in a database for a specific crime without exposing all of the other people listed in the database to unwarranted surveillance? How can a technology user confidently wear a smartwatch or install a health-tracking app when a future court decision might turn their digital medical records into evidence against them?

These are just a few of the many analogous challenges we'll explore in this book. Though they encompass a broad range of settings, contexts, technologies, and cultures, what unites them is a phenomenon that we refer to as *the secret life of data*.[7] The basic premise is simple, though its implications are complex:

> There is no limit to the amount and variety of data—and, ultimately, knowledge—that may be produced from an object, event, or interaction, given enough time, distance, and computational power.

In even plainer language, it means that whatever we think we're sharing when we upload a selfie, write an email, shop online, stream a video, look up driving directions, track our sleep, "like" a post, write a book, or spit into a test tube, that's only the tip of the proverbial iceberg. Both the artifacts we produce intentionally and the data traces we leave in our wake as we go about our daily lives can—and likely will—be recorded, archived, analyzed, combined, and cross-referenced with other data and used to generate new forms of knowledge without our awareness or consent. This may be done not just once, but over and over, by individuals and institutions we've never heard of, using techniques that may not even have been invented yet, as technology and society continue to coevolve over time.

All of this adds up to something greater than the sum of its parts. Billions of us now live in a globally networked world, and our social structures and personal relationships reflect this fact in innumerable ways. This means every time any of us creates or collects a piece of data, even through casual interactions with everyday technologies, we have a responsibility to consider how the information derived from it might affect our friends, our colleagues, our compatriots, and billions of people we'll never even meet. To borrow a phrase from a popular twentieth-century bumper sticker, we need to "think globally" whenever we "act locally."

(Throughout this book we will use the terms *data*, *information*, and *knowledge* a great many times. We do not treat these terms interchangeably, and we use them very deliberately. *Data* refers to discrete, quantitative values extracted from larger, more complex phenomena.[8] *Information* refers to those mathematical values put into context and thereby given a social meaning. *Knowledge* refers to actionable inferences made from information by interested parties such as governments, marketers, or researchers. This flow from data to information to knowledge is, in our current society, an industrialized process, and many of the technologies and businesses we discuss are part of that pipeline.)

It's a lot to wrap your head around. Some readers may find themselves thinking something like, "Holy smokes! I'm flushing my smartphone down the toilet immediately and moving to a cabin in the woods." Others might respond with something closer to, "Big deal, who cares? I stopped believing in privacy decades ago."

Both of these reactions are completely understandable, but neither is particularly constructive. For one thing, it's not just about smartphones or even the internet; the secret life of data applies to *every* technology and behavior that may leave a trace, from walking past a CCTV camera on the street to borrowing a library book to registering to vote. For another thing, it's not just about privacy. The secret life of data adds new complexities and challenges to every aspect of our lives, from our identities to our personal and professional relationships to our democracies. It's impossible to escape and dangerous to ignore. That's why this book suggests a third possible reaction: "I understand what's at stake, and I feel empowered to help shape the future of technology and society."

Toward that end, we have attempted to cut through the hype and hand-wringing that pervades much of the popular discussion of technology and its role in public life. Instead, we hope to provide you with a window into how things arrived at this point, a sense of who the different actors and stakeholders are, and a guide to recognizing their hallmarks in the technology you interact with every day.

INVITING A WIDER CONVERSATION

There are already many other voices weighing in on the relationship between technology and society with insights that range from sublime to ridiculous and prognostications that range from dire to downright utopian. While some of these discussions tailored to a general readership are rooted in sound scholarship and lived experience, many are sensationalistic or overtly partisan and don't hold water under closer scrutiny.

Conversely, much of the scholarship on technology and society that we admire dives deeply into a specific aspect of the larger issue—a promising or dangerous new invention, an important new approach to law and regulation, or a threat disproportionately faced by a particular group of people. The singular focus of these books and articles, addressed to expert readers, can often provide a boost for much-needed policy change or spur a new direction in technological investment and development.

Unfortunately, these two communities of readers rarely interact productively. General interest texts tend to skirt thornier problems and technical details, while deeper discussions may be couched in complex terms that feel exclusionary to casual readers. This makes it difficult for many people to participate in conversations about the future of technology and society, even though these issues will bear heavily on all of our futures.

That's why our aim is to bring a wider swath of voices into dialogue with one another by charting a middle path between these approaches. We draw on the theoretical perspectives and research findings of the academic community, and our conclusions are based on both reasoned analysis and personal commitments to human dignity and social justice. But we're not trying to move the needle on a specific policy or to advance the bleeding edge of any particular field. Instead, we use plain language, straightforward storytelling, and dozens of expert interviews to help demystify issues that are far too often shrouded in tech jargon and policy wonkiness to help readers make sense of the many profoundly weird, troubling, and exciting changes wrought by technology in our lives and relationships.

We acknowledge we can't accomplish this on our own because, like anyone else, our worldviews are limited by our personal experiences and identities. We are two middle-aged, cisgender, Ashkenazi Jewish men from New York City who happen to have been close friends since we met on the first day of ninth grade at a public high school focused on math and science. Our shared interests in speculative fiction, musical and visual cultures, political activism, and the transformative potential of new technology brought us together as teenagers, and this book is one manifestation of the lifelong conversation we've been having on those themes since then. Along the way, we've each gained many of the experiences and credentials necessary to write a book like this—higher educational degrees; jobs in tech, marketing, and design; creative professional profiles; stints as department chairs at university programs focused on tech and society; and some coauthored academic journal articles—but the range of subjects we discuss and how we play with ideas haven't changed much in all that time.

If our conversation has evolved since our youth, it has as much to do with the maturing of digital technology itself as it does with our own aging. When we first explored online services like Gopher and Archie in the early 1990s, it was via text-based terminals in our university computer labs. Taking inspiration from early cyberpunk novels like William Gibson's *Neuromancer* and Neal Stephenson's *Snow Crash*, we imagined a future in which data were ubiquitous and the lines between physical space and information space, human and machine, became increasingly porous. Despite the cautionary elements in both these books, we were excited at the prospect of an always-on, digitally connected society. Like many people in our age cohort, we spent countless hours fantasizing about what our adult lives in cyberspace might be like.

As young adults, we both worked in the technology sector, principally as researchers focused on how the dot-com boom of the late '90s would transform education, cultures, and economies by linking together billions of individuals and institutions on a global scale. We understood implicitly that our youth and enthusiasm for tech were crucial qualifications for the job. Unattached as we were to legacy industries and the old ways of doing business, we were free to imagine the many ways in which the internet might accelerate change or improve upon the status quo. And we were rewarded both economically and reputationally for our ability to translate these utopian visions to our elders, the decision makers at large corporations, federal agencies, nonprofit organizations, and other seats of power.

During the Web 2.0 years in the first decade or so of the twenty-first century (when we were in our thirties), we ceased thinking about digital technology principally as a *potential* change agent, and approached it instead as a *practical* one. We began academic careers, producing scholarship and teaching classes about new creative models like "remix culture," new legal models like "copyleft," new "real-time" software environments for interactive art, and new collaborative platforms like social media. Our job functions shifted away from translating the future to our elders, and moved toward exploring present opportunities and hazards with our contemporaries. As creative professionals and technology builders, we also took advantage of new media production tools to advance our own powers of expression, putting theory into practice and building collaborative networks with far-flung colleagues across the globe.

In the mid-2010s, as we entered our forties, we began to cast a warier eye on digital networks and data industries. The tech sector became increasingly consolidated, with a handful of companies (Google, Apple, Facebook, and Amazon) collecting the vast majority of data and dollars associated with people's online—and, increasingly, offline—activities. At the same time, geopolitical shifts including a widespread authoritarian turn brought into stark relief the political stakes of living in a digitally connected world. In both measurable and ineffable ways, the optimistic, freewheeling spirit of the early internet years was giving way to a grimmer, more utilitarian approach to technology.[9] For our part, as teachers, artists, and researchers, we found that our roles were shifting once again, to explaining the predigital past to people too young to have lived through it.

We offer this capsule joint biography not in order to hog the spotlight but to give you a better sense of what our strengths and weaknesses might be in analyzing and describing the secret life of data. We benefit from having enjoyed a front-row seat to some of the biggest shifts in technology and society during our lifetimes, but we also suffer from our long-standing investment in a vision of a more egalitarian datafied society that is at odds with the current reality, both in our native United States and elsewhere around the world. We're researchers and scholars, but we're also artists and idealists, and some of the impetus for writing this book comes from our discomfort with how the social consequences of digital technology have diverged so drastically from our hopes and ambitions.

We are also hobbled, as all social researchers are, by the fact that we are attempting to describe large-scale, long-term phenomena from a fixed point (technically, two fixed points that are not so distant from one another). There are regions, identities, moments, and perspectives to which we simply lack direct access and that we can only learn about second- or thirdhand. We have done our best to account for this by reaching out to experts across a range of nations, ethnicities, gender identities, professional roles, and political perspectives and including their voices directly in the pages that follow. We interviewed twenty-nine such experts, most of whom agreed to speak on the record, over the course of writing this book, and solicited feedback on the manuscript from roughly a dozen more. We also had many more informal interactions with friends, colleagues, and acquaintances while presenting or discussing our ideas at conferences, colloquia, and community events. We couldn't have written this book without their insightful contributions and critiques, and we are grateful to each of them for what they've taught us.

ABOUT THIS BOOK

The Secret Life of Data is organized thematically, beginning with some basic principles and then expanding its scope outward from their more immediate and intimate consequences to their larger-scale ramifications. Each chapter builds on the conclusions of previous chapters, but also stands on its own as an investigation of a particular set of challenges and opportunities related to data and society. Each chapter introduces new case studies, new expert voices, and new conceptual tools. And yet, because all aspects of our subject matter are by definition interlinked and interdependent, the book has many recurring themes. A discussion of biometric devices would be incomplete without acknowledging the role of data brokers, a discussion of political surveillance would be incomplete without acknowledging the role of biometrics, and so forth. By the time you finish reading this book, our hope is that you will have an overarching sense of how these diverse technological and human networks fit together.

We begin the book with one of the most straightforward examples of data hiding in plain sight: metadata. In chapter 1, we discuss the ancient origins of practices like tagging, data classification, and cataloguing, and we demonstrate how these techniques, originally designed for libraries of papyrus scrolls, have been integrated into modern digital storage formats and databases. We then explore the role of metadata as a vector for unintended revelations, from forensic investigations to mass surveillance. We also discuss the human toll of such revelations, interviewing the sister of whistleblower Reality Winner. Winner was an intelligence analyst whose role in leaking information about Russian interference in the 2016 US presidential election was exposed through metadata, possibly contributing to her conviction for violating the Espionage Act and her five-year prison sentence.

In chapter 2, we move beyond metadata and address an even larger and less well-defined reservoir for the secret life of data: what's typically referred to by computer scientists as *unstructured data*. Broadly speaking, this term refers to all the stuff that wasn't designed or intended to be collected and analyzed using computers but ends up stored in databases anyway. This includes a wide range of items, from ancient relics to vintage media artifacts to digital ephemera like text messages and draft documents. We also discuss some of the downstream consequences of analyzing unstructured data, such as turning the absence of data into its own form of useful information and using machine learning algorithms to make highly educated guesses about missing data (for instance, taking blurry photos and digitally sharpening them).

After discussing many of the ways in which data might turn up where it's least expected, we shift our focus to some of the social consequences of these secret lives, addressing the perennial question: What could go wrong? In chapter 3, we talk about how today's highly datafied societies differ from those of previous eras and how these changes undermine long-standing values and expectations like anonymity, provenance, and creative attribution. We interview technology researchers Mutale Nkonde and Mar Hicks to explore the ubiquitous challenge of algorithmic bias, in which preexisting social hierarchies and abusive power relations are coded (even unintentionally) into the algorithms that are used to make important decisions about our lives. We also interview public-health specialist Emily Tseng, who describes how the secret life of data may be used abusively in interpersonal relationships in contexts such as cyberstalking.

After addressing the secret life of data in broad strokes, we home in on specific aspects of the issue, beginning with the role of "smart" devices in our lives. In chapter 4, we interview internet governance scholar Laura DeNardis about the challenges posed by cyberphysical systems like the Internet of Things (IoT). We also speak with computer security researcher Vasilis Mavroudis about the consequences of living in a social environment where many of the objects we interact with on a daily basis collect data about us using digital sensors and networking technology. Finally, we discuss the data broker industry that monetizes and trades the data collected by cyberphysical systems by interviewing Dennis Crowley, the founder and former CEO of Foursquare, a major broker that collects and sells geographic tracking data about pedestrian shoppers.

In chapter 5, we delve into even more intimate territory, discussing the role of biometric sensors (technology that measures, analyzes, and tracks the human body). We explore the benefits of such technologies, like life-saving medical interventions, as well as some of the dangers, including the increasingly central role of facial-recognition software in public spaces, law enforcement, and political oppression. We interview Sam Gregory, a director at human rights organization Witness, to discuss some other consequences of biometric data analysis, such as *deepfakes*—convincingly manipulated videos that make it seem as though a person is doing or saying things they never did or said. Other interviewees include Maggie Clifford, a graduate student who discovered a previously unknown sibling thanks to a genealogy website like the ones used to unmask the Golden State Killer, and Sharrona Pearl, a medical ethicist who helps us trace the connections between nineteenth-century pseudosciences like physiognomy and some of the more questionable applications of biometrics in the twenty-first century.

Next, we explore some of the deeper social, cultural, and psychological consequences of biometrics and cyberphysical systems. In chapter 6, we begin by discussing the quantified self (QS) movement, a trend in the 2010s of people using digital sensors and diagnostic tools for the purpose of self-knowledge and self-improvement. We also discuss the flip side of this trend—the increasing evidence that the unchecked use of such technologies can have damaging psychological implications, especially for vulnerable people, like adolescents. We also interview design researcher and disability rights advocate Laura Forlano, who explains how some biometric technologies intended to have beneficial medical effects can have cataclysmic health consequences instead if disabled users are left out of the design and development process. Finally, we introduce the concept of *algo-vision*—the widespread and disorienting experience of seeing oneself through the "eyes" of an algorithm—and we discuss the potential benefits and dangers of a world in which millions of people have internalized this algorithmic gaze.

In chapter 7, we expand our scope again, to explore global-scale sensor networks and the social implications of integrating the entire physical world, from micro to macro scales, into the technological stack. We discuss what we call *crowdsourced stewardship* in the form of environmental surveillance initiatives. We also introduce the concept of "triangulation" as a model for artificial intelligence and machine learning systems. Triangulation is based on a multiperspectival approach to knowledge creation as a more equitable and accurate alternative to the single-point perspectives presented by many platforms. We also discuss how adding the physical world to the digital stack introduces new conduits for the secret life of data (such as Easter eggs), and new security vulnerabilities (such as malware). Finally, we discuss the promise, peril, and hype surrounding the metaverse and mixed reality and ask whether the benefits of these immersive digital platforms are worth the potential cost of privatizing public space by inundating it with a deluge of proprietary data.

In our final chapter, we consider the implications of the secret life of data for the future of democratic society. We discuss several troubling trends and developments from the digital Cold War, characterized by ongoing multinational cyberespionage and hacking campaigns, to *neurotargeting*, a dangerous new form of propaganda that uses the cutting-edge tools of digital marketing to bombard emotionally volatile citizens with calculated disinformation. We interview information warfare scholar Emma Briant, one of the world's leading experts on the Cambridge Analytica scandal, to explore how these techniques were used in the 2016 US presidential election and how they are still being refined and used around the globe nearly a decade later. We also interview criminal defense attorney Paul Hetznecker, who describes how *predictive policing*—an algorithmically assisted approach to law enforcement that targets suspects prior to any criminal acts—poses major risks to civil rights. Finally, we offer a few glimmers of hope: new ideas, technologies, policies, and trends that we believe can help preserve democratic norms and institutions in a data-rich society.

We conclude the book with a broad evaluation of the social, cultural, and institutional consequences of the secret life of data, and we discuss a range of new ethical, regulatory, and technological approaches that may help to harness the power of data without allowing its secret life to run amok. We also introduce new experts, such as Yves-Alexandre de Montjoye, a professor of computational privacy. His insights help explain why there's no such thing as a quick fix to these entrenched challenges and why, if we value social justice, inclusive cultures, and democratic governance, we must use every tool at our disposal to shape the future of technology and society by shifting our focus away from the mechanisms of data extraction and analysis and toward the larger social contexts in which these processes unfold.

In our final pages, we consider potential change agents looming on the horizon, such as the quantum cliff and "general" (human-level or greater) AI. Technological leaps such as these may not only change the course of the secret life of data but also undermine the solutions we've developed to counteract its more damaging consequences—potentially wreaking havoc on our intimate lives and civic institutions. On the other hand, we still know so little about these potential threats that we can't yet see their full implications, which means that we must all remain vigilant at the vanguard of new technological and social developments. That's why we end the book by encouraging our readers to play an active role in shaping the future—using your common sense, professional expertise, political power, and cultural influence to critique new technologies as they emerge and to advocate for a networked society that centers human agency and dignity.

We encourage you to keep this in the back of your mind, as you read this book. With each new technology we introduce, each new expert voice we add to the chorus, and each unexpected consequence we review, ask yourself, How does this affect my own life? What does this mean for the values I hold dear? And, most importantly, What can I add to this conversation?

We'll be listening. And you can be certain that others will, as well.

1 DATA ABOUT DATA (ABOUT DATA)

Even if you're not a history buff, you've probably heard of the Library of Alexandria. It was built in the third century BCE, while Mediterranean thinkers like Euclid and Archimedes were working on radical new scientific concepts like geometry and engineering. The library was first conceived by Alexander the Great, both as a testament to the size and power of his empire and as a means of collecting the wealth of knowledge from all the lands he'd conquered, translating them into Greek and thus Hellenizing his dominion under a common language and cultural identity.

Though Alexander died before his library was built, his successors, Ptolemy I and Ptolemy II, followed through on his plan, amassing one of the greatest troves of recorded knowledge in the ancient world. In order to manage this collection, they hired the Homeric scholar Zenodotus to serve as its first superintendent.

The library housed roughly half a million scrolls on subjects ranging from science to law to philosophy to literature. Mostly written on papyrus, the scrolls were in perpetual danger of decay, and each had to be copied by hand and replaced periodically in order to preserve the knowledge it contained. It was Zenodotus's job to ensure not only that this cultural archive was preserved for the benefit of future generations but that contemporary scholars could find the information they needed and contribute their own work to it, as well.

In order to accomplish this Herculean task, Zenodotus developed several techniques that are still in use today. One of his principal innovations was to standardize some of the older texts in the library—for instance, reconciling different versions of Homer's classic epics *The Iliad* and *The Odyssey* and creating a single, consistent version of each. Another technique was to classify the scrolls into what today we would call different genres, storing each set of related texts in its own room—dedicating one area solely to medical texts, for example, and another to poetry. He also introduced the handy idea of alphabetization, so it was easy to sort through an entire collection of scrolls and to figure out where to put one back once it was no longer in use.

To make it even easier for librarians and scholars to keep track of things, Zenodotus decided to add tags to each scroll with information such as the author and title of the work so that the documents wouldn't need to be unrolled and rerolled each time someone wanted to know what they contained. Finally, Zenodotus had a catalog written in which all of the works the library held were listed using the classification, alphabetization, and tagging techniques described above.

If Zenodotus's innovations as a librarian seem like common sense to those of us living in the twenty-first century, it's only because they were so successful that, over time, they became standard practices for anyone trying to keep track of large collections of information. In fact, we now have a handy term that covers the classification, ordering, tagging, and cataloguing techniques developed at the Library of Alexandria: *metadata*.

To use the simplest definition, metadata is data about data. It tells us where a piece of information came from, where it belongs, what it contains, how to use it, and any number of other helpful things we might want to know before we actually unroll a scroll, search for a book in the library stacks, or open a digital file. Metadata has helped humankind to develop a geometrically expanding store of scientific and cultural knowledge for millennia. Without it, it's unlikely modern history would have unfolded in the way it did, and it's hard to imagine historical milestones like the Renaissance, the Industrial Revolution, or the development of the internet.

Yet, although Zenodotus's innovations made it easier for people to store, access, and contribute to large archives of information, the development of metadata also introduced some new challenges that have bedeviled librarians and those of us who rely on their valuable work ever since. If you've ever typed a URL into a web browser and gotten a "404 page not found" error or spent an hour on the phone with a customer service representative trying to clear up a billing error, you know this firsthand.

First of all, metadata aren't always accurate. And since metadata are easier to reproduce than the larger texts they describe, errors can end up replicating like a virus. If you accidentally type "Johm Coltrane" into the "artist" field of an MP3, for instance, good luck ever getting one of your digital music players to correctly credit the song to John Coltrane. And these kinds of small mistakes can have ripple effects with more serious consequences. By some estimates, metadata errors such as these have already cost musicians billions of dollars in lost royalties.[1]

Another challenge is one that Zenodotus probably didn't see coming: Metadata are generative. On their own, they are intended to serve as a kind of shortcut, describing a larger piece of data as concisely as possible. But together, metadata can add up to more than the sum of their parts, taking on new meanings and functions and spawning even more metadata in turn. To use an old-fashioned example, a phone number is a piece of metadata about you. It's a short string of digits that tells people how to call or text you. But thousands of phone numbers assembled together become a new object with a new set of social functions: a phone book. And there are other metadata about phone books themselves: who publishes them, what geographic areas they cover, how many pages they have, who counts as the head of a household, and so forth. So metadata beget more metadata, ad infinitum. And that means that Zenodotus's fantasy of cataloguing all human knowledge can never be fulfilled, because the act of cataloguing creates more categories of knowledge—a scenario whose tragic consequences are famously lampooned in Jorge Luis Borges's dark, satirical short story "The Library of Babel."[2]

Finally, even though they generate new meanings and contexts for knowledge, metadata can also be destructive. The standardization of Greek texts in the Library of Alexandria was designed to erase and obsolesce other versions of the books it contained, including versions in other languages, in order to unify Alexander's empire. Along similar lines, today's metadata generally privilege one version of a text over another. And in doing so, they privilege one person or community over another, flattening out the diverse and multivocal cultural landscape and erasing the voices and perspectives of the less powerful. This can serve as an instrument of larger power dynamics. To use an extreme example, when Nazi Germany started stamping the letter *J* on the passports of Jewish citizens in 1938, this new form of metadata both reflected and enforced the political subjugation of Jews throughout the country while normalizing the racist logic behind the policy.

METADATA IN THE MODERN WORLD

In the two millennia since Zenodotus first arrived at the Library of Alexandria, metadata have evolved considerably. The techniques he developed, including standardization, classification, alphabetization, tagging, and cataloguing, are still in widespread use today, but they have expanded to fit the growing range of creative platforms, storage technologies, cultural styles, and social uses for information. Think back to the last time you watched a movie, read a book, streamed some music, or even used an app on your phone to buy food or meet a date. Whether you used the Dewey decimal system or swiped right, you were relying on metadata to discover exactly what (or whom) you were looking for.

Today, just as in ancient Egypt, people around the globe use metadata to help navigate the world of knowledge and information. But it's no longer just about finding the right reading material; now, we use metadata to organize our countries and cities, to manage the flow of commodities and communications through the world's supply chains and information networks, and to structure our intimate lives and interpersonal relationships. Metadata is, in short, fundamental to the architecture of modern industrialized life in the twenty-first century.

As the uses for metadata have proliferated, so have its political and social consequences. Because metadata is *generative*, it always has the capacity to add more to the story, rather than merely summarizing something longer and more complex. Because it's *destructive*, metadata always erases as it creates, reinforcing one perspective while sidelining another. And because it doesn't exist in a vacuum but features in a broader landscape of data and information, metadata can be *triangulated* with other datasets to produce unforeseen outcomes and sometimes startling revelations. In other words, one of the first places to start looking for the secret life of data is in metadata. The vast accumulation of metadata is like a river that runs right beneath the surface of our daily lives, carrying a torrent of potential new cultural meanings and implications, never more than a faucet or a flood away from coming to light.

If this premise sounds familiar to you, it's probably because you've seen the social consequences of metadata discussed in the news over the past decade or so (even if it wasn't discussed in precisely these terms). Public awareness of metadata spiked in 2013, when a US government contractor named Edward Snowden leaked information to British newspaper *The Guardian* about several secret National Security Agency (NSA) surveillance programs.

While Snowden's leaks were, and still are, the subject of political controversy, they revealed a widespread pattern of unregulated data collection by the government that fundamentally challenged our assumptions of privacy in the twenty-first century. Snowden's documents showed that the NSA had for a decade collected "bulk internet metadata," including the addresses of email senders and recipients as well as internet users' IP addresses (unique identifiers for individual machines and local networks connected to the internet). At the same time, the agency had archived and analyzed all of the "telephony metadata" listing the phone numbers, call duration, and cell phone tower locations of every call made on the networks of US phone carrier Verizon. The scope of metadata collection included millions of Americans and countless foreign citizens on a daily basis.

After these secret programs were revealed, the US government claimed that they posed no threat to everyday Americans' privacy or security, in part because they weren't capturing or analyzing the contents of the emails or calls themselves. After all, the NSA argued, it was "only metadata."[3] Who cared that the government knew that person X had called person Y on Thanksgiving, or that someone at company A emailed someone else at company B last Tuesday?

As it turns out, a lot of people cared. A Pew Research study found that in the six months following Snowden's revelations, the percentage of Americans who said they approved of government metadata collection programs dropped by one-fifth, and public opinion regarding these programs flipped from a majority approving to a majority disapproving of them.[4]

In the meantime, civil liberties advocates and privacy researchers began to document the ways in which the kinds of information the NSA collected amounted to more than merely the sum of their data points. Computer scientists from Stanford University reproduced the NSA's methodology in miniature, collecting supposedly anonymized smartphone metadata from 823 participants and doing their best to analyze the quarter of a million phone calls and 1.2 million text messages those participants sent and received over the course of a few months.[5]

Their findings were unequivocal. First, despite the NSA's claims, it was trivial for someone with enough processing power and coding skill to deanonymize telephone numbers, connecting them and their metadata to real human beings and businesses. Second, it was relatively easy to locate those people and businesses in the physical world, even without additional information such as GPS records. Third, it was pretty clear what kinds of relationships existed between the people in the dataset; for instance, married couples stuck out from the rest of the participants like a sore thumb because of their unique communication patterns. Finally, it was very easy to infer sensitive information about deanonymized users—including their health status, religion, drug use, and whether they owned weapons or were seeking an abortion—based on the other phone numbers they interacted with.

If that sounds like a shocking amount of personal information to derive from a few months' worth of anonymized phone records of 823 random people, remember that it's only a microscopic fleck in the galaxy of knowledge the NSA can derive from its database, which contains trillions of phone calls, emails, and web visits made each year. As journalist Barton Gellman has detailed in his extensive research on this program, the NSA didn't merely collect metadata about the communications of hundreds of millions of Americans; it created a secret system that used Facebook-like technology to assemble those metadata into a "social graph" with highly detailed profiles of nearly everyone in the country and a map of the connections between them spanning years of contact.[6] With a single search query, analysts using this tool could infer innumerable things about the intimate lives and personal histories of American citizens that their own families and friends might not be privy to (and, for that matter, that may not even be accurate).

The NSA's surveillance program is a far cry from the tags hanging off the ends of ancient scrolls, but the implications of metadata aren't as different in the two scenarios as it may appear at first glance. In both cases, the metadata were created for one purpose but ended up being used to generate new meanings by people far removed from their original contexts. The creators and stewards of the Library of Alexandria used metadata to Hellenize the Mediterranean, imposing the politics of empire on the diverse peoples of the region; the NSA's surveillance program served its own government's geopolitical agenda, turning an individual's web of personal associations into a metric gauging their Americanness or their otherness—a dividing line that grew in potency and power in the wake of the attacks of September 11, 2001.

Yet there are notable differences between the two cases as well. Most importantly, in the Great Library, only the scrolls were subject to this coding, analysis, and knowledge creation. The NSA applied these techniques to people, or at least to the public record of their private lives. And this program is only one example of the rapidly expanding range of metadata collection programs undertaken on a global scale by governments, businesses, criminal enterprises, and others without the consent—or even the knowledge, in most cases—of those being tracked.

What does it mean for us as a society that so much of our daily activity, from the objects we make and consume to the messages we share with our families, friends, and associates, generates an ever-expanding volume of metadata, revealing far more about us than we can keep track of, let alone understand? What bits of our lives are included and excluded, how are they categorized, and how does that shape other people's understanding of us as individuals and communities? Who is collecting and analyzing these metadata, and for what purposes? What kinds of unexpected uses have they already found? What might other people, in other times and places, make of them? Finally, what mechanisms do we have, if any, to discover and correct false inferences that are made about our personal lives, choices, and allegiances?

These questions are too complex and numerous to answer fully, and we will spend the rest of this book grappling with them from various perspectives.

But for the remainder of this chapter, we will analyze some important examples of how metadata serve as a conduit for the secret life of data. The consequences include revealing hidden strings of political power, forging chains of liability and provenance, guiding investments in technology and culture, and even consigning government whistleblowers to prison. Though there's much more at work in these stories than a few bits of metadata coming to light, none of them would be possible, for better and for worse, without the techniques Zenodotus first developed to serve Alexandria's library.

THE METADATA IS THE MESSAGE

One of the reasons metadata can be so revealing is that there is no single source or format for it. Even a lone digital file can generate multiple overlapping and potentially contradictory layers of information when placed in new contexts. For instance, the manuscript for this book is currently being drafted in Microsoft Word, which has metadata fields built into its file structure. These fields contain information such as who authored the document, when it was created and modified, how many revisions have been made, and so forth. Most people don't ever look at the metadata generated by the documents we write (to be fair, why would we?), but it's not exactly classified information. In Word, all you have to do is choose the Properties option in the File menu to view and even edit some of the metadata about your document.

But that's only the first layer. This manuscript is stored on a hard drive on a Mac laptop computer, and the MacOS operating system creates its own metadata about the document. Because the book is coauthored, we also keep a draft of the manuscript on the Google Docs cloud-based collaboration platform, which creates its own set of metadata, including copies of every draft and a detailed history of who accessed and edited the document, when, and how. We will also share this manuscript with other people, such as our editors and colleagues; when we do, the emails we send will generate their own metadata about the attached document. And when the file is saved on their computers, their operating systems and word processors will create additional metadata about it. Remember, all of these different layers of metadata describe the exact same set of words—the ones you're reading right now.

These metadata are often duplicative. If we send you a copy of our manuscript, the word count field will be the same for you as it is for us. But frequently other metadata will differ. The information about who accessed the file and when will be different. You may live in a different time zone, or speak a different language, or use a different operating system, in which case your computer will have to translate our metadata to match your preferred formats. Little differences like this can reveal a lot. In a way, metadata for a computer file tell you as much about the end user as they do about the contents or author of the file itself. Or, as Marshall McLuhan might have said, the metadata is the message.

METADATA AND THE (UN)OFFICIAL STORY

Most of the media we produce in our personal and professional lives, like emails, photos, and written documents, generate metadata automatically, often without our knowledge, participation, or consent. These metadata always tell a story to those who know how to listen—and sometimes that story conflicts with the one that the document's author is trying to tell.

In addition to the obvious privacy issues they raise, these conflicts, inconsistencies, and irregularities can also reveal much bigger tensions in our social, political, and cultural systems, including relationships of power and influence that were intended to remain hidden. This can be a great benefit for truth seekers like researchers and investigative journalists, but it may also pose a challenge for those who prefer to operate away from the public spotlight for personal, political, or professional reasons. An illuminating example is the use of metadata in exposing lobbyists' surreptitious role in crafting legislation and policy.

In theory, the role of lobbyists is to inform and educate lawmakers. For instance, a legislator may come to office with expertise in foreign policy but know little about health care. Another congressperson may be an expert in agriculture but have scant experience when it comes to infrastructure. That's where lobbyists come in. They help lawmakers understand important policy issues, but they do so from a perspective that benefits their paying clients. So the senator who knows nothing about health care might learn about the wonders of government-run insurance from a lobbyist retained by a labor union, while the congressperson who needs to understand infrastructure better might get an earful about the need for new highways from a lobbyist paid by a big contractor. These messages often come along with a fancy dinner, a trip to an exotic locale, or even help with fundraising, making the lawmaker potentially more receptive to the lobbyist's recommendations.

This may sound unethical, but it's not illegal. It's a well-established and heavily regulated practice. In fact, in the United States, lobbyists are required to leave a fairly detailed paper trail about who pays them and how that money is spent, and those records are freely available to the public on websites hosted by organizations like Open Secrets and FollowTheMoney.org.

But one thing lobbyists aren't supposed to do is write laws and policy themselves. That task falls solely to the people who were elected or appointed to serve their constituents. If the highest bidder got to set the rules that we all must live by, they might end up tilting the scales of justice so much that nobody else could get a fair shake. So it's important that government officials remain independent from lobbyists and their clients, making policy in the public interest, rather than just rubber-stamping the agendas of powerful special interests.

Unfortunately, that's not always how it works. Whether because of laziness or corruption, or simply by mistake, sometimes public officials who claim to be creating laws and policies from scratch are apparently passing off a lobbyist's words, verbatim, as their own. How do we know this? In many cases, it is because journalists have discovered these hidden connections in the metadata embedded in lawmakers' own documents. To list just a few examples:

- In 2004, California attorney general Bill Lockyer drafted a letter to his fellow state attorneys general decrying the dangers of peer-to-peer file-sharing software and advocating for a new law requiring software developers to add digital warning labels to their products or face prosecution for fraud. *Wired* examined the metadata in a leaked Microsoft

Word draft of the letter and found that the Author field was labeled "stevensonv"—presumably indicating Vans Stevenson, the Motion Picture Association of America's vice president for state legislative affairs.[7] According to public records, Lockyer received nearly a quarter of a million dollars in contributions from the TV and film industry during his two terms as attorney general.

- In 2013, thirty-two Democratic congresspeople signed a letter to the Department of Labor opposing new regulations that would have increased protections of Americans' retirement accounts from exploitative speculation by dishonest investment brokers. Though the letter claimed their opposition was based on concerns that the regulation could "severely limit access to low cost investment advice," *Mother Jones* analyzed the document's metadata and found that it listed Robert Lewis, a lobbyist for the financial services industry, as its author. According to analysis by *Mother Jones*, the lawmakers had received "tens of thousands of dollars in campaign money from the securities and investment industry."[8]

- In the 2014–15 Congressional term, Senator Lindsey Graham and Congressman Jason Chaffetz introduced several bills that aimed to make nearly all forms of online gambling illegal. Their support for the bills was framed as a moral intervention, tailored to protect Americans from what Graham called "nefarious activity" that would benefit "people in the criminal world and terrorist world."[9] But political news site *The Hill* obtained a draft copy of the legislation and found that its metadata showed its author was Darryl Nirenberg, a lobbyist for the Las Vegas Sands casino, then owned by political megadonor Sheldon Adelson.[10] This strongly suggests that the true purpose of the bills wasn't to protect Americans from nefarious activity but rather to protect a powerful political patron from online competition.

The number of similar incidents is beyond counting and hardly surprising. Political watchdogs have known for centuries that lobbyists can play a central role in shaping legislation and policy, potentially undermining the public interest. What's new in these recent cases is the role that metadata play as forensic threads for investigative journalists to pull. This new technique ups the ante for both lobbyists and reporters, creating new challenges for those who would rather operate in secret while also providing new avenues of inquiry for those in the business of shining light on shadowy practices. It also serves as an excellent reminder of how the secret life of data connects us all, whether we want it to or not: one person's actions (like forwarding a document without scrubbing the metadata fields) can have consequences for someone else's life (like inviting public ridicule or winning a Pulitzer).

EVERY CONTACT LEAVES A TRACE

In the summer of 1922, the bustling French town of Lyon had a mystery on its hands: someone was rifling through the mail and stealing bills, checks, and other valuables from sealed envelopes. The postal department suspected an inside job, so they conducted the investigation themselves instead of immediately alerting the police.

A clever agent figured out that the dirty deed was probably being done in the post office bathroom, so he hid in the room above it and peeped down through a hole in the ceiling. Sure enough, a postal clerk showed up in the loo with a stack of letters and methodically opened them one at a time, pocketing the valuables secreted within. Unfortunately, the inspector couldn't identify the culprit by looking through a hole at the back and top of his head, so he decided to mark the clerk by stomping on the ceiling and dislodging some plaster dust onto his uniform. Hearing the knocking sound, the clerk beat a hasty retreat, but too late to escape detection. The police were alerted, and when the suspect's coat was analyzed under a microscope, it revealed traces of plaster dust and thus confirmed his guilt.

To those of us with fifteen seasons of CSI available for instantaneous streaming on our phones and tablets, this might sound like a pretty ho-hum case. But a century ago, the idea of conducting police investigations like this was a radical new idea. In fact, the perp made the unfortunate decision to commit his crime in the same city as the world's very first forensic police laboratory, which was founded by Edmond Locard, often referred to as the Sherlock Holmes of France.

Locard—who credited Holmes as an inspiration—is perhaps best known for pioneering the use of fingerprinting in police investigations. But, like his fictional mentor, his interests were far broader. Aided by new developments in microscope technology, Locard believed that close investigation of features invisible to the naked eye, such as dust, pollen, and textile fibers, would allow savvy investigators to reconstruct the events of the past and connect people to the scenes of their crimes even without witnesses or traditional forms of evidence. This idea was crystallized in what is now referred to as Locard's exchange principle, which proposes that "every contact leaves a trace." In other words, two objects can't interact without each being altered by the other, and if we learn how to observe and analyze those alterations, we can demonstrate that the interaction occurred.

So what does this all have to do with metadata and the secret life of data? Everything. Even in the digital world, where no plaster dust falls from the ceiling, no seat cushion fibers stick to miscreants' pants, and no tobacco ash smolders near a victim's body, every virtual contact still leaves a trace. And that means other people, armed with the right forensic tools, can always track your metaphorical digital fingerprints back to you. Even the most careful cybercriminals can still sometimes get caught this way, just like the savviest cat burglars can be identified by a single strand of errant DNA found at the scene of a crime. And no matter how careful we think we're being, there's no telling what innovative forensic technologies might emerge to reveal the traces of our past. A century ago, Locard was inspired by advancements in microscopy. Today, the latest forensic innovations are driven by algorithms and artificial intelligence.

Of course, most websites and apps aren't exactly Fort Knox, and most internet users are a far cry from elite hackers, which means it doesn't take a forensic innovator like Locard to connect us to the digital traces we leave behind. In fact, a lot of digital sleuthing is now done by amateurs on the internet rather than professionals in police labs. And metadata fields are some of the first places they start looking for such traces.

Unless you're a professional photographer or software coder, you may not have heard of EXIF, short for exchangeable image file format. It's the metadata embedded in a digital image file, like a JPG, TIF, or PNG, whenever you take a photo. When digital cameras were first introduced, EXIF data were used to store information about the photos themselves, making it easier for photographers and editors to sort and process the newly massive volume of images that could be produced by a single user of a single device. But once digital cameras were integrated into mobile phones with internet connections, EXIF data expanded to encompass more and more information about the person taking the photo, including when, where, and how the image was captured.

Today, photos taken by many popular mobile devices, such as the iPhone, include geotagging information by default in their EXIF headers. This means that, unless you've disabled this setting, every time you share a selfie via social media, text, or email, you're also sharing your precise GPS location at the time you took that photo.

EXIF metadata is therefore frequently used by journalists, law enforcement agencies, private investigators, and others to connect human beings to both the digital images they create and the physical places they visit. This is how one of the most infamous fugitives in recent years, John McAfee, was caught.

Ironically, McAfee made his name and his fortune as a cybersecurity entrepreneur, founding the popular antivirus and identity protection company that still bears his name (although he resigned from the organization in 1994). In 2012, McAfee was living in Belize when he was identified as a person of interest in the murder of his neighbor, an American expatriate named Gregory Faull, who had complained to the police that McAfee's "vicious dogs," armed security guards, and traffic "at all hours of the night" were a threat to the community's safety and peace of mind.[11] Though not officially a murder suspect, McAfee fled the area. Yet his seemingly insatiable need for attention led him to start a fugitive blog on which he taunted the Belizean authorities and invited journalists to join him on the lam.

One of those journalists, photographer Robert King, took a picture of McAfee standing next to *Vice* editor-in-chief Rocco Castoro in front of some jungle fronds. *Vice* published the photo online with a story headlined "We Are with John McAfee Right Now, Suckers."[12] Less than two hours later, a computer security researcher named Mark Loveless revealed that the *Vice* journalists were, in fact, the suckers, tweeting "Check the metadata in the photo. Oooops. . . ."[13]

The image illustrating the article still contained the photo's original EXIF data, generated by Castoro's iPhone camera, which revealed the precise GPS coordinates where it had been taken. Those coordinates, as we have ascertained through our own analysis of the metadata, were adjacent to the Hacienda Tijax hotel in Rio Dulce, Guatemala, demonstrating that McAfee had crossed the border to escape the Belizean police. Two days later, McAfee was arrested in Guatemala for entering the country illegally. A week after that, he was deported to the United States.[14]

EXIF geotags aren't the only kind of metadata that can land people in trouble for their surreptitious behavior, and megalomaniac fugitives aren't the only people who risk being outed by their digital fingerprints. Another prominent recent example is someone who not only shunned the limelight and the world of crime but actually tried to serve the public interest by blowing the whistle on her own government when she believed it had failed to protect its citizens.

In the spring of 2017, the United States and the world at large were still reeling from the unexpected results of the recent presidential election, in which the openly bigoted, transparently self-serving television personality Donald Trump defeated the consensus front-runner, former secretary of state Hillary Clinton. Though there was widespread speculation during and after the campaign that the Russian government was meddling in the election to help improve Trump's odds of winning, the US government—including the president himself, vociferously and repeatedly—continued to deny any direct knowledge of such meddling (as did Russian president Vladimir Putin).

That's why Reality Winner, an idealistic twenty-five-year-old Air Force veteran from South Texas who worked for a federal contractor processing intercepted foreign communications for the NSA, was shocked to discover what seemed like direct evidence to the contrary. At work one morning, she saw an NSA memo, available to all employees with top-secret clearance, detailing the agency's assessment that state-sponsored Russian hackers had attacked a US voting software supplier and sent targeted spear-phishing emails to hundreds of local election officials containing malware that would make their computers even more vulnerable to further hacking.

As her sister Brittany told us, Reality's immediate response was, "Well, this is the proof that everybody's been waiting for. This is what everybody deserves to see." And though sharing a top-secret memo with unauthorized people is a felony under US law, Winner printed it out on her office printer, folded it up, put it in an envelope, and mailed it from a street corner postal box to investigative journalism outlet the *Intercept*.

"She felt that it was her civic duty," Brittany explained. "She felt like the American people deserve to know about this. Our taxpayer money is being used to fund these people who put together these [secret] reports. So why is there information that . . . people with a [security] clearance can see, but not the average American person, who wasn't really sure whether or not this meddling had occurred? She wanted that information to be out there."

As an intelligence professional, Reality Winner believed she was covering her tracks well. There was nothing visible on the printed memo or the envelope she sent it in that could be traced back to her directly, and she expected that the *Intercept* would serve as a firewall of sorts, offering her the same kind of anonymizing protection afforded to historical whistleblowers like Deep Throat, a crucial source for the story of President Nixon's role in the 1972 Watergate break-in.

Nonetheless, on June 3, two days before the *Intercept* published its story about the memo, the FBI showed up at Winner's home and arrested her. She was held without bail until her trial. A year after her arrest she accepted a plea deal on a single charge of espionage, for which she received a sentence of over five years in prison and three years of supervised release.[15]

While there is much in Winner's story to concern scholars of information and technology, we tell it here because, like the more salacious tale of John McAfee, metadata may have played a central role in her capture. Nobody has revealed publicly how the FBI determined that Winner was the whistleblower, and there are several popular theories, ranging from the mundane to the sinister. As Brittany Winner told us, Reality's own family remains uncertain on this point, though they have their suspicions. One thing we do know is that, following standard journalistic practice, the *Intercept* verified the memo's authenticity before publishing or reporting on it. They did this by scanning a copy of the paper document to PDF format and emailing it to the NSA to give the agency the opportunity to deny its veracity (it never has). Though the scanning process would have removed many of the traces that Locard's forensic lab relied on (such as fingerprints, dust, and textile fibers), it preserved, and even amplified, other ones.

One such trace was the presence of printer tracking dots—minuscule yellow marks printed on nearly every page of paper by several of the most popular commercial laser printers currently in use. Though human eyes can't detect them, these dots, which serve as a kind of barcode revealing the printer's serial number and other metadata about the printed document, are clearly visible to digital cameras. So if you scan a printed page containing tracking dots to a digital format and adjust the color balance of the resulting image file, the dots will stand out in stark relief, overlaid in a pattern across the entire page.

Shortly after the *Intercept* published its article on the NSA memo, on the heels of Winner's arrest, several internet sleuths pointed out that the PDF of the memo, which the *Intercept* posted in full to its website, contains tracking dots and therefore identifies both the printer and the precise time at which the memo was printed out.[16] We verified this ourselves, downloading the PDF and adjusting the color balance of the file to produce the image shown in figure 1.1. Using a tracking dot decoding guide published by the Electronic Frontier Foundation, we ascertained that the NSA memo scan was printed at 6:20 a.m. on a Xerox DocuPrint model printer with serial number 29535218.[17]

We're merely scholars and artists, not a crack team of cyberdetectives. So if it was this easy for us to uncover the metadata that connected the PDF online to the time and place where Reality Winner printed the NSA memo, it would have been a no-brainer for the NSA to do the same when they were asked to verify the document's veracity.

This is hardly conclusive proof that the *Intercept* was responsible for revealing Winner's identity as the whistleblower to the federal government. There are many other traces, digital and otherwise, that may have connected her to the document she mailed. As Micah Lee, the *Intercept*'s director of

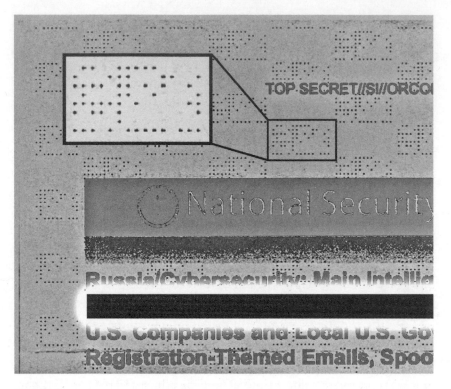

Figure 1.1
An excerpt of the NSA memo adjusted to reveal the secret tracking dots on the page.

information security, told us in an interview, Winner's employer was almost certainly keeping internal computer logs tracking exactly who accessed, edited, or printed the files on their servers, and the organization would have shared those logs with the NSA if requested. There were also likely surveillance cameras covering the mailbox where Winner dropped the envelope. And because tracking dots were never mentioned in the search warrants and affidavits filed against Winner, Lee said he does not see "evidence suggest[ing] that the leak investigators at the FBI even knew about" the existence of such dots.

Regardless of whether the dots played a role in Winner's story (and, for that matter, whether you sympathize with her actions), their existence on the incriminating PDF is a potent reminder of the new risks we face as an ever-greater volume of metadata is imprinted on the physical and digital artifacts we interact with daily—often without our knowledge. It also creates new ethical challenges for the institutions and individuals who serve as go-betweens. Journalists have always had to balance the competing priorities of protecting their sources and verifying the integrity of their evidence, but when you add new technologies to the mix, that balance can easily be tipped too far in one direction or another. To put it another way, the *Intercept*'s practices might have been good enough to protect a source like Deep Throat in the context of a 1970s technological environment, but they hadn't evolved sufficiently to ensure Winner's anonymity.

To their credit, staff members at the *Intercept* have acknowledged this lapse and taken steps to correct it. As Lee told us, when Winner sent them the printed memo, "We didn't really have any policies to make sure that we had a security person involved" in verifying the document while protecting its source. "So, in the aftermath of Reality Winner, we revamped all that stuff. And so now we have actual policies and guidance in place to make sure that things don't fall to the ground."

For her part, Brittany Winner said she welcomes any policy changes at the outlet that would "protect their sources better in the future." Yet she also acknowledged that "there's nothing really more that they can do to help Reality. . . . They can't get her out of jail."[18] In the absence of such a miracle, she said that at the very least she'd appreciate some acknowledgment of negligence on the part of the *Intercept*: "An old-fashioned apology, I guess, would be great."

METADATA: THE NEW BUSINESS AS USUAL

We have discussed several ways in which metadata shines a light on the secret life of data: revealing hidden influence of lobbyists on lawmakers; pinpointing the location of fugitives; and linking top-secret government documents to whistleblowers. Yet it's important to remember that, though the details of these particular incidents may be embarrassing, unexpected, or even shocking, the general fact of their existence is hardly surprising. Metadata aren't an accidental byproduct of computer processing or an esoteric form of code reserved for the likes of elite hackers. They're a commonplace, everyday feature of mainstream digital technology that were designed intentionally to be just hidden enough so that most people don't bother to think about them.

Although the stories we've told in this chapter are on the sensationalistic side (for your reading pleasure), more mundane versions of these events happen by the millions, all the time, around the world. Metadata are part of business as usual now—in government, in commerce, in cultural expression, and in our interpersonal relationships.

Though metadata were originally created to help organize libraries of content, they have now been absorbed into the process of content production itself. And they are no longer merely a description of the content in question; they're continually updated to reflect how audiences interact with that content. To put it another way, in the twenty-first century, every movie, song, document, or photo is likely to be accompanied by a historical account of how it has been used, by whom, and under what circumstances. Consequently, tracking this information is increasingly valuable to governments and corporations when they're making big decisions about how to invest their resources or adjust their policies. Entire cottage industries have evolved to serve these needs. If you've ever wondered what "data analytics" or "business intelligence" firms do, now you know.

We spoke to an executive at a major streaming video company who walked us through the company's process to choose whether to produce a new television show or movie and whether to renew or cancel an existing one. In the old days, such decisions would be based on the gut instincts of studio chiefs or the golden ears of record label execs. Today, the streaming executive told us, "Everything that we do is data-driven. Programming decisions are based purely on viewing habits."

If someone pitches a new science-fictional film about time travel, for instance, the streaming service would "take all the data from all the sci-fi movies—all the time travel movies in particular—and based on existing metrics" such as box office gross, popularity on streaming services, and more granular data such as completion rate (the ratio of the number of people starting a show to the number of people finishing it), the company will "predict how well it would do, and that determines how much money we can spend on the movie, where it would be valuable to us as an asset."

Executives in the streaming-media industry see this as a snowballing process: "the more content we make, the more accurate the data becomes," and therefore, they believe, the more likely the company is to see a return on its investment (ROI).[19]

By some measures, this strategy has been successful for data-driven entertainment companies; for instance, Netflix chief content officer Ted Sarandos told investors in 2017 that 93 percent of the company's original series had been renewed, compared to only about one-third of series at traditional television networks.[20] Whether this success in generating ROI is equally successful as an engine of cultural progress is another question altogether. If you've ever felt like Netflix or Hulu keeps recommending different versions of the same show over and over, or if you've lamented the lack of diversity among the casts of these shows, you can thank the company for relying heavily on an algorithm engineered to minimize risk instead of prioritizing experimentation.

As the value of metadata has expanded, so has its volume. Not only are there geometrically more metadata-enhanced documents with every passing year, but the number of metadata fields associated with each document and the amount of information populating each of those fields have continued to grow as well. Depending on your measuring stick, the volume of metadata may far exceed the volume of actual data now, potentially by orders of magnitude. Of course, this is a difficult claim to support, because, as we have demonstrated, metadata has a habit of becoming data, and generating more metadata about itself so it's difficult to draw a clear dividing line between the two categories. If you want to get philosophical about it, you could say there's no longer any intrinsic difference between data and metadata—as we view the world through a computational lens, every piece of data we produce is, by definition, data about data. And, in a society increasingly reliant on computer networks for the continuation of everything from our economy to our culture to our democratic institutions, the structures and categories we use for our metadata come to determine the structures and categories we use to understand one another and ourselves.

This explosion of metadata isn't necessarily a good thing or a bad thing. There are clear social benefits to it. Streaming media companies use it to improve their ROI and fine-tune their recommendation algorithms. Projects like the Content Authenticity Initiative (a partnership launched by Adobe, Twitter, and the *New York Times*) are developing systems that use metadata to improve public trust in journalism by verifying the sources of articles and images shared on social media. Even the ubiquitous tracking dots used by laser printers were developed for an important reason: to help detect counterfeit currency and prevent its circulation.

Yet there are many social costs, as well. As Harvard-based security technologist Bruce Schneier put it when Snowden's leaks were first published, "Metadata equals surveillance data."[21] The precipitous rise in metadata generation and collection hasn't been balanced by a rise in public awareness, let alone a comprehensive shift in policy empowering us to prevent governments, corporations, and even-less-savory actors from peering ever deeper into our personal and professional lives.

Part of the problem is that the incentives to produce, collect, and analyze metadata are so great that virtually no social cost can outweigh the perceived benefits to either the surveillants (who profit) or the surveilled (who enjoy the convenience of a well-organized digital universe). Part of the problem is sheer momentum; as internet governance scholar Laura DeNardis has pointed out, our global network of automated data collection and processing systems would continue chugging along for years even if every human being on the planet disappeared overnight.[22] And part of the problem is another variety of momentum: metadata fields are grandfathered into new technology based on our old ways of doing business. When *Vice* posted the photo of John McAfee to accompany their reportage, they were doing the same thing newspapers had done since the Civil War. But their photos contained EXIF data, which wasn't a problem for journalists covering the siege of Petersburg.

To make things even more complex, this cascading torrent of metadata doesn't exist in a vacuum; it's part of an even larger sea of data and information that structures our societies and permeates our cultures. This increases the power of metadata, because metadata can be used in conjunction with other data to corroborate information. Tracking dots may have been used to identify Reality Winner as the NSA whistleblower, or they may have been used to verify what was already known from the internal server logs at the government contractor where she worked or the surveillance footage from the mailbox where she sent the memo to the *Intercept*. Looking at metadata is a great way to spot the hidden fingers of lobbyists guiding lawmakers' work, but it's only one of many. Publicly available tools like the Legislative Influence Detector, hosted by the Data Science for Social Good Foundation and Carnegie Mellon University, achieve similar ends by looking for substantial similarities between the texts of lobbyist documents and policy documents, using text-matching techniques similar to antiplagiarism software.

Furthermore, because it's the new business as usual, metadata has become another tool for powerful institutions and individuals to maintain their power and social standing. Whoever gets to choose the categories we fit into, how those categories are scored, what sources of data are used to score them, and which data they'll be compared to has a tremendous amount of influence over how we end up interpreting and using those data. Facebook has famously come under fire in recent years for bundling together groups of users to sell to advertisers, relying on emergent categories like "Jew haters" and "interested in pseudoscience" to enable anyone with a big enough budget to promote disinformation to millions of social-media users.[23] While these practices may strike us as abhorrent, they're hardly unusual; the entire internet essentially runs on targeted advertising, which itself is entirely reliant on unaccountable and frequently opaque systems of metadata collection and analysis.

Ultimately, we must understand that, though metadata are an important conduit for the secret life of data, they are also just one element of a much larger and more convoluted landscape that encompasses virtually every aspect of our lives. What seems random and chaotic today might be codified as data tomorrow and start generating actionable metadata the day after that. There's no telling how our tools and imaginations will continue to evolve as computer processing grows in power and scope. But, as with Zenodotus's innovations, we can be certain that they will continue to generate more knowledge even as they overwrite and obsolesce older ways of knowing, streamlining and autotuning the cacophony of human voices into a standardized symphony that swells under the batons of our algorithmic conductors.

2 ALL DATA ARE BIG DATA

The difference between *objective* and *subjective* is something many of us are taught while we're still in middle school. One, we're told, is unbiased, fair, rational, scientific, *trustworthy*. The other is biased, selfish, emotional, cultural, *untrustworthy*. By the time we're adults, many people in industrialized societies have internalized this distinction, and it subtly shapes how we view the world and the people around us. Most relevantly, it shapes how we interact with digital technology; because computers are presented as fundamentally objective calculating machines, we are encouraged to trust them and to invite them into our private lives and public spaces.

But are these machines really worthy of our trust? Does their calculating nature really mean that they're unbiased? Not necessarily. The digital devices and online services that many of us use in our daily lives rely heavily on data collection techniques and statistical analytical tools that, despite their appearance of objectivity, can mask some very subjective judgments. This contradiction is not unique to twenty-first-century computers or global data networks. In fact, if we examine the social history of statistical analysis, it's very clear that, from the beginning, the line between subjective and objective has been blurry at best.

One of the founders of modern statistics was a man named Francis Galton, a cousin of Charles Darwin. He is also frequently remembered today as the father of eugenics, the racist, misogynist, pseudoscientific principle that human beings could—and should—selectively reproduce (by choice or by force) to improve traits such as intelligence, morality, and beauty.[1]

For Galton, there was little difference between statistics and eugenics. His methods justified his conclusions and vice versa; he used the language of objectivity to rationalize his own worldview. One clear example of this is the way he developed his statistical techniques in the first place.

As Galton related proudly in his autobiography, he combined his interests in heredity and statistical analysis to construct a "Beauty-Map" of Britain. Using a cross-shaped piece of paper and a "needle mounted as a pricker" hidden in his pocket, Galton evaluated the desirability of the "girls" he passed on the street. "Attractive" women would merit a pinprick at the top of the paper cross, "repellent" ones would get a prick at the bottom, and "indifferent" ones would be marked in the wider middle.[2] Once he assembled the data, Galton compared the averages between different cities in the British isles, claiming that London had the most beautiful women and Aberdeen the fewest.

Galton's "Beauty-Map" wasn't constructed merely to objectify women, though it both reflected and rationalized his paternalistic attitudes. He also wanted to demonstrate what he believed was an important scientific principle: namely, that human biology (and, by extension, society) followed the same kinds of probabilistic rules as other aspects of the natural world, such as the patterns of values shown by a pair of repeatedly tossed dice. To Galton, the distribution of pinpricks on his papers—most clustered near the middle with relatively few holes at the top and bottom—were a physical representation of the normal distribution curve (sometimes referred to as a bell curve), which has indeed been observed in a broad range of phenomena, such as the birth weight of newborn babies, the surface temperature of the Earth, the height of trees in a forest, the amount of milk produced by dairy cows, and the size of snowflakes.[3]

As repugnant as Galton's ideas may seem today, it's hard to overstate how influential they were, in Britain and around the world, in his time and throughout the following century. At the most extreme, his theories of eugenics were used to justify atrocities from Nazi mass executions to South African apartheid to the forced sterilization of untold thousands of poor and nonwhite women in the United States.[4]

Yet Galton's "Beauty-Map," in its own way, was even more influential than his calls for "Race Improvement." The premise that human beings can be understood, both individually and collectively, through data collection and mathematical analysis became one of the core tenets of modern society, and its impacts can be seen across a broad range of institutions and cultures around the globe. The concepts of a "normal" person, an "average" household, and a "top-rated" policy or product owe not merely their philosophical origins but the very language we use to describe them to Galton's biased vision of the world. To this day, the premise of statistical objectivity serves to justify, amplify, and obscure a multitude of subjective opinions, often in the exercise of social power.

IMPOSING ORDER ON CHAOS

We have already discussed metadata and the ballooning volume of information that accompanies every photo, video, and document we produce in the digital era. Yet metadata constitutes a tiny foothill compared to the mountains of data being produced and collected by the trillions of sensors, processors, and databases currently operating on our planet. If recent trends continue, the volume will continue to grow geometrically in the years to come.

Metadata is an example of what computer scientists call *structured data*. Each bit that is generated, such as a GPS coordinate or the name of an author, is ready-made to fit a preexisting category in a database. In other words, computers can be programmed to "know" what to do with the information as soon as it's produced. But much of the data collected by our expanding global network of sensors and platforms, from surveillance camera footage to the content of our social media posts, begins as *unstructured data*: it doesn't conform neatly to a preexisting set of categories, and therefore nobody is quite sure what to do with it at first.

Turning unstructured data into structured data (and using algorithms to make it knowable, useful, and therefore valuable) is one of the growth industries of our era and one of the defining forms of social power in twenty-first-century society. Yet the process is anything but straightforward; companies like Amazon, Facebook, and Google are racing to capitalize on this opportunity, but they're not all running toward the same finish line. That's because the kind of knowledge they produce depends as much on how they structure and process data as it does on what they've collected to begin with. And those decisions are themselves based on their underlying agendas: what they think is interesting, important, or potentially profitable.

Think of it this way: When Galton was strolling the streets of Britain with a pin in his pocket, he was busy giving structure to unstructured data. But the results of his analysis depended more on what he was looking *for* than whom he was looking *at*. If Galton had been scoring passersby based on their height or their apparent weight instead of his own assessment of their attractiveness, he may have confirmed his normal distribution hypothesis in a slightly less disturbing way—and he wouldn't have had the same opportunity to validate his subjective, paternalistic opinions about women by giving them the veneer of objectivity.

Furthermore, if Galton had been open to other kinds of observations—how happy someone looked, or how many different shades of green they were wearing, or whether they smelled like tobacco—he may have discovered entirely different kinds of truths about his environment that had less to do with supposedly universal statistical patterns and more to do with the social realities governing the lives of the people around him.

There are a multitude of ways in which people today, frequently aided by cutting-edge digital technology, are still following in Galton's footsteps, whether they know it or not. New technologies are turning unstructured data into structured data and helping to make sense of the cacophony that surrounds us, but they are also attempting to impose order, value, judgment, and most importantly, *power* over innumerable aspects of the human experience.

Not only do these tools and technologies reflect the biases and agendas of the people who built them, but they are also principally used by the privileged to serve their own interests. This has profound implications for the shape of our society in the years to come and for the role that data plays in struggles for justice and equity. Even the most benevolent or banal cases we describe in this chapter have the capacity to shift the balance of social power and destabilize long-standing practices, beliefs, and organizational systems. And that means that whenever we hear about some new, gee-whiz technology that purports to benefit the world in some way, we have to do the difficult work of asking ourselves which parties are actually benefiting and at whose expense.

TOOLS OF THE TRADE

The tools and technologies we'll discuss in this chapter might sound diverse and weird enough to fill the shelves of a sci-fi warehouse, but remember, as you read, that they fit into just a handful of categories with well-defined functions:

- New *sensors* capable of collecting forms of previously inaccessible data.
- New *forensic techniques* capable of identifying previously overlooked data.
- New *algorithms* capable of detecting previously hidden patterns in data. These algorithms are often created in order to take advantage of new sensors and forensic techniques. Like Galton's paper crosses, they are built to identify features that the programmers think are important or interesting.
- New *artificial intelligence (AI) and machine learning (ML) models* capable of taking patterns extrapolated from a given dataset (sometimes referred to as training data) and making decisions, with varying degrees of confidence, in other contexts.
- New *generative tools* capable of using AI and ML models to produce new data that are predicted by trends identified in preexisting datasets.

If these terms and concepts are unfamiliar to you, don't worry. Even people who work in the fields of media and technology can find the dizzying pace of innovation overwhelming and sometimes get lost in the digital weeds. You can always refer back to this section for a quick refresher. But we hope it won't come to that—we'll do our best to explain what these tools actually do and how they work as we discuss their many uses and abuses.

What's most important to remember is that, at a fundamental level, all of these systems are statistical in nature. They deal in probabilities, not certainties; their outputs are more akin to educated guesses than incontrovertible facts. For instance, sensors like cameras, microphones, and environmental gauges have well-known and highly detailed sampling errors. Manufacturers typically report in their technical specifications that these devices are only accurate to a certain point and can't be relied on beyond that margin of error. Along similar lines, forensic techniques are merely efforts to reconstruct what *might* have happened; it's up to their human users to determine the most likely interpretation of the results. Pattern-spotting algorithms are engaged in guesswork, as well—just like human beings have a tendency to see connections where they don't necessarily exist (such as perceiving a butterfly in an inkblot), these processes have a bias toward connecting dots, whether these connections are warranted or not.

AI and ML models can be understood in this light as guesswork built upon guesswork. In addition to relying on the potentially flawed results of sensors, forensic techniques, and algorithms, they add a new layer of uncertainty through their reliance on feature extraction. This means that a large set of training data is reduced in size and complexity by focusing on key elements that either the programmers believe are important (like eyes, mouths, and noses in digital photos) or that stick out mathematically (like an unusually hot or cold day in a year's worth of weather data). While this makes it much easier for AI and ML models to make sense of large volumes of data, it also introduces new errors and biases by giving some data more weight than others. As the output of these models appear more seamless and "realistic," the models' biases become harder and harder for human observers to spot.

Figure 2.1
Feature extraction performed on an AI-generated face (analyzed using TouchDesigner).

REVIVING RELICS

In 2020, a team of undergraduate students from the Rochester Institute of Technology made a fascinating discovery. By shining ultraviolet light on a fifteenth-century Christian text called the *Book of Hours*, they discovered a hidden layer of writing beneath it—a "dark French cursive" text that had nothing to do with the religious content printed over it.[5]

While this might sound like a plot point from the Harry Potter universe, there was nothing remotely magical or supernatural about it. The page was a palimpsest—a piece of parchment that had been scraped clean and recycled by monks working on an illuminated manuscript. Before the printing press was invented, this was a common practice developed to make the most of the limited supply of writing materials and labor.

Once the original French text was scraped away, its traces were invisible to the naked eye, and the monks who printed the *Book of Hours* probably assumed that was the end of it. Neither they, nor the author whose work they erased, could possibly have known that, six hundred years in the future, a bunch of students would rediscover the lost text with the aid of electromagnetic radiation imaging technology.

Recent advancements in both imaging and artificial intelligence have contributed to an explosion of discoveries like this one and a renewed interest in what critics and historians call *media archaeology*—the reinvestigation of "dead" media and the cultural ideas encoded in them. Together, these discoveries are not only reshaping our understanding of the past but also forcing us to confront our relationship to the future by anticipating the secret afterlives of the media we produce today.

In a way, every piece of recorded data—in any medium and from any age—is like a palimpsest. Even digital files that have supposedly been deleted from hard drives usually leave invisible traces that can be recovered easily with the right tools. There's no telling what technological or cultural developments may allow future investigators to shine new light on our own media, revealing something long hidden, presumed invisible, or even captured unknowingly in the first place.

One of the most common applications of these new techniques is in the field of art history. Newspaper stories appear almost daily about some cultural artifact that has been given new life or fresh meaning with the aid of innovative technology. In 2019, for instance, the archaeologist Cecilie Brøns released a digital reconstruction of a nearly two-thousand-year-old Syrian bust known as the Beauty of Palmyra.[6] A combination of high-resolution visual sensors and AI were used to reconstitute the missing fragments of the sculpture's nose, lips, and hand and to identify trace elements of the pigments once used to decorate its surface. While the original portrait, first excavated in 1928, remains chipped and colorless, like most ancient artifacts, Brøns's digital reconstruction shows a complete figure decorated in vibrant colors that *might* look far more familiar to a second-century Syrian than the surviving relic.[7]

Along similar lines, researchers at the Lawrence Berkeley National Laboratory have developed a novel technique to preserve old audio recordings.[8] Nineteenth-century records, which contain a wealth of irreplaceable material (including spoken examples of dead and dying languages, the voices of historical public figures, and long-forgotten musical styles), are often too fragile to play even once on a phonograph, which has made their contents inaccessible for a century. The new technology, called IRENE, uses ultra-high-resolution imaging to make a virtual model of the crumbling wax cylinders and shellac discs, then "plays" those virtual discs with a virtual phonograph in order to create a digital audio reconstruction of the contents. It's kind of the opposite of a 3D printer: instead of turning a virtual model into a physical object, IRENE takes a real-world object and lets you play (with) it in simulated virtual space. Of course, the sound it produces is only a well-informed guess; because AI models are probabilistic, there's a small but genuine chance that its inferences will be wrong. And because the original objects are too delicate to play back, there's no way to verify whether they're right.

Sometimes, these techniques are used not merely to refresh or reconstruct a cultural artifact but to discover elements that may have been invisible in the original, finished artwork. Like medieval palimpsests, oil paintings contain many hidden layers, such as early drafts of the compositions or even older paintings that were covered over when a canvas was reused. These layers have recently become accessible thanks to new forensic technologies.

In recent years, art historians have learned that Rembrandt's *An Old Man in Military Costume*, currently in the collection of the Getty Center, obscures an older, upside-down image of a man wearing a cloak.[9] And Pablo Picasso's Blue Period portrait *La Miséreuse accroupie* is painted sideways over a mountainous landscape, originally produced by an unknown Spanish artist, in a way that cleverly repurposes elements of the sideways landscape in the backdrop to Picasso's newer figure.[10] These kinds of revelations are not merely fascinating in their own right; they also offer art historians a deeper understanding of the social, material, and cultural environments in which art was made, and therefore a deeper understanding of the artwork itself.

There are even instances in which these techniques are used to reveal things intentionally left hidden or ambiguous by the original artists. For decades, art historians speculated about the identity of the person who defaced Edvard Munch's expressionist masterpiece *The Scream* by scrawling "can only have been painted by a madman" in pencil over the image. Recently, using infrared imaging and handwriting analysis, curators at the National Museum of Norway demonstrated that the painter himself was most likely the "vandal," perhaps defacing his own work in a fit of pique.[11]

Using different techniques to similar effect, two statistics professors teamed up with a music industry executive to create a machine learning computer model that analyzes Beatles songs attributed to the songwriting team of John Lennon and Paul McCartney and predicts which pieces were written by which of the two celebrated composers. Like all such models, this one deals in probabilities rather than certainties, boasting "greater than 75% accuracy" in its predictive power. And, like the virtual phonograph records in the IRENE project, there's no way to verify its accuracy (at least, not without asking McCartney, whose own memories might be biased or blurred by time).[12]

These new tools and techniques have applications far beyond these spheres and implications that carry more weight than the answer to who really wrote "Carry That Weight" (it was Paul). A team of archivists and scientists recently demonstrated a technique that combines an imaging technology called X-ray microtomography with computational flattening algorithms to probe the contents of ancient, sealed letters without damaging or even touching the paper they're written on.[13] Like the IRENE project, this will allow historians to access materials long believed to be inaccessible to modern researchers, while preserving the material artifacts of antiquity. There are thousands of such letters in collections ranging from the Vatican to university libraries, and their contents will undoubtedly contribute some new chapters and debates to the field of world history.

Digital forensic techniques have already yielded some historical revelations that intersect with today's political and social struggles in unexpected ways. New analysis of a photograph of nineteenth-century literary titan Joaquim Maria Machado de Assis, considered by many to be the greatest writer in Brazil's history, showed that the portrait had been artificially lightened. For a century Machado had been misrepresented, through this alteration of what had been his de facto official portrait, as "a blond guy with blue eyes," in the words of Brazilian professor José Vicente, who led the reconstruction effort.[14] The new, digitally restored image portrays Machado with medium-complexioned brown skin that suggests to many modern observers he had African heritage. In the context of racist and colorist biases in contemporary Brazilian society originating in the era of the Atlantic slave trade and European colonization of the Americas, the reassessment of Machado's appearance resulted in controversy and challenged the racial hierarchies that continue to devalue Black artists and citizens in Brazil to this day.

THE NEVER ENDING STORY

When the classic sci-fi film *Blade Runner* was first released, in 1982, its dystopian portrait of Los Angeles in 2019 seemed incredibly futuristic and yet all too plausible. The vision of a smog-choked megacity populated by artificially intelligent criminals and lorded over by a technology titan in his gleaming tower served both as a promise of the marvels of the dawning silicon age, and as a potent warning of how our growing reliance on computers might undermine our fundamental humanity.

Once the actual year 2019 rolled around, sci-fi fans and tech pundits alike had a great time dissecting the film on the internet: what it got right, what it got wrong, and what might yet be possible, for better and for worse.

While sentient androids and flying cars are still the stuff of futuristic fiction, some of the film's predictions have panned out. We indeed have building-sized video billboards, smartphones, voice interfaces for computers, and, of course, evil tech titans.

Yet while flashy gadgets and gizmos, both real and imagined, may have captivated the attention of most *Blade Runner* fans, a certain subset of geeks who watched the film found their hearts racing fastest when Decker, the protagonist, used a piece of fictional tech called the Esper machine. This device did what every librarian, media scholar, and data scientist knew beyond a shadow of doubt was impossible: it allowed the user to look more deeply into a photograph, discovering details in the image that were not apparent on its surface.

The promise of the Esper was universally tantalizing to those who collect and organize data for a living; nothing is more frustrating than reaching a dead end in a dataset, reading the last word in a memoir, or squinting to see beyond the maximum resolution in a blurry photo. Yet information science told us unequivocally that the Esper could never be more than a pipe dream. The camera only captures what it captures; the book ends, and that's all she wrote, as the saying goes. As it turns out, however, the Esper might not be science fiction, after all. New sensors, techniques, and AI models are helping forensic scientists, artists, and everyday technology users do what was considered impossible not very long ago: create "new" knowledge from a finite artifact. Examples range from deconstructing old musical recordings to mapping light waves in a physical space to synthesizing sound from silent video.

You've probably heard a fair number of mashups. It's a style of remixed electronic music that uses a piece of one song (usually a vocal track) and a piece of another (usually an instrumental track) to make a new song that is, ideally, greater than the sum of its parts.[15] Mashups were a big deal back in the first decade of the twenty-first century, when the idea of making music on a laptop computer still seemed new and exciting. In the 2020s, they're old hat, produced routinely for advertisements, promotions, and awards shows.

Commercial mashups like these are made with the blessings of record companies, who will send a producer raw "stems"—isolated vocal and instrumental tracks—making it relatively simple to mix them together with crystal clarity. It's a bit more difficult for amateur mashup artists (or *bedroom producers*, as they're sometimes known). Because they often can't get isolated tracks directly from the source, they've developed a range of tricks and techniques to pare down commercially mixed recordings, dampening the bits of sound they don't want and emphasizing the bits they do. While the resulting mashups are often impressive, they're never perfectly free of unwanted sonic artifacts and can sound muddy or distorted. That's because, like paint, recorded sound is easy to mix but impossible to unmix.

Or rather, it *was* impossible. Today, a new set of machine learning tools allows music producers to dive into a fully mixed sound recording, isolate a single instrument from the mix, and pull it out whole, like a fish from water. As with all AI models, this process is probabilistic, meaning that sometimes it gets things wrong. Usually, though, the resulting track sounds more or less identical to the original stem. This technology has given rise to a whole new set of commercial possibilities based around the concept of *upmixing*—taking old, premixed recordings, separating them into their constituent stems, and producing shiny, new, multitrack mixes for movies, games, and other contexts. The technology is even being used to turn classic recordings into karaoke tracks for people who want to sing along with the original instrumental accompaniment. And, naturally, bedroom producers are having a field day, using open-source demixing software like Spleeter and free or cheap laptop and tablet apps like VirtualDJ and Moises to create mashups and remixes that sound nearly as clear as commercially produced work based on the original stems.

Another new AI-assisted forensic technique is called *structured light*. Though it has many different variations and applications, they're all based on the same basic principle. Light waves, arranged in a known pattern, are projected from an illuminated source or sources. Then they do what light waves do—they bounce around and hit things, reflecting off of surfaces and distorting the original pattern. What structured light technology allows you to do is reverse engineer this process. Because we know what the light looked like when it was first projected, we can build a probabilistic model of what happened to that light from the moment of projection until its detection nanoseconds later by a digital camera. The engineer and artist

Kyle McDonald, for instance, has demonstrated a technique that allows him to construct a usable 3D map of a room and its contents just by looking at the patterns of light reflected by a bunch of disco balls onto their surfaces.[16] The surveillance powers of this technology haven't gone unnoticed; in recent years, military and intelligence researchers have experimented with using structured light as well, to "see around walls" and provide tactical advantages on the battlefield and in other combat situations.[17]

While upmixers can deconstruct and reconstruct sound, and structured light can add new dimensions to visual information, there are also novel and futuristic techniques that create a kind of technological synesthesia, converting information from one medium into another in much the same way that certain people "taste" music or "hear" colors.

For instance, researchers at MIT used a technology called Eulerian video magnification to build what they call a "visual microphone."[18] Basically, the technology can synthesize sonic information from a silent video clip, analyzing the minuscule vibrations on the surface of an object, such as a plastic bag, a piece of aluminum foil, or the leaf of a tree, and reading those vibrations as audio much like the IRENE project probabilistically converts 3D scans of old records into sound waves. It has also become increasingly simple for computer vision programs to recognize text in libraries of photos and videos, using a fairly old form of technology called optical character recognition (OCR). Other computer vision algorithms can recognize and tag discrete objects like clothing, tools, plants, and animals. These examples of technological synesthesia aren't just cool; they're potentially life-changing for a broad variety of end users, increasing the accessibility of information stored in both old and new media for audiences who may be hard of hearing, have low vision, or have other disabilities.

The social impacts of these new technologies actually go far beyond what *Blade Runner* imagined for the Esper device. That's because, like so many science-fictional narratives predating the 1990s, the film didn't anticipate the consequences of a globally networked society. Thanks to the internet, these new forms of digital forensic enhancement aren't limited to a single user examining a single piece of media. Instead, such techniques, in part because they are now so widely available, can analyze entire libraries of content comprising millions of photos, videos, and other pieces of recorded media. The Flim.ai search engine currently allows paying users to search 751,000 high-definition film stills for elements including clothing and scenery. Google's GDELT project offers a free search interface for global television, radio, print, and web news archives that includes textual searching of video captions and features real-time translation between sixty-five languages.[19] And recent versions of Apple's consumer-grade photo app allow users to search for and interact with text in images in their own libraries, such as dialing a phone number printed on a store awning in the background of a years-old snapshot.

In other words, the entire internet is becoming an Esper machine, and every piece of digital media will continue to yield greater and greater volumes of knowledge as these forensic techniques continue to develop and interconnect. We can't predict these specific techniques—or their consequences—any more than the creators of *Blade Runner* could see the internet coming. But we can be certain that what seems today like "all she wrote" will just be the prologue of the story someday soon.

FAKE IT WHILE YOU MAKE IT

The digital techniques we've described so far are mostly forensic in nature; they use improvements in sensor technology like cameras and microphones as well as increasingly sophisticated AI to generate new knowledge from and about old media. Sometimes they get it wrong, of course, but the aim is to get it right. But that's not the only potential use for these new capabilities. They can also be used to create things that never existed in the first place—either augmenting an existing piece of media or creating one from scratch.

Augmentation is everywhere these days. Early popular applications tended to be playful, like Snapchat filters, which could add laser eyes, rainbow vomit, or a pair of horns to a selfie. This wasn't very challenging computationally, because any two people would vomit an identical rainbow; all the AI had to do was identify someone's mouth and add the necessary animation. But more recent applications are more computationally intensive, and for that reason, more useful. For instance, media production software company Adobe recently released a feature in their popular Photoshop app called Super Resolution that does just what the name suggests—it increases the resolution of an image, turning a ten-megapixel photo (for instance) into a forty-megapixel photo. It does this not by discovering hidden information and restoring it to the file, but by extrapolation and inference. Adobe trained several ML models on low- and high-resolution versions of millions of different images, and the resulting algorithms help to predict the most likely appearance of any low-res image at a higher resolution.

Along the same lines, artists and technologists around the world have been busy developing algorithms to automatically "up-rez" and colorize old black-and-white photos and videos. Director Peter Jackson even used these techniques for an entire feature-length documentary about World War I called *They Shall Not Grow Old* (2018). If and when three-dimensional video becomes a popular entertainment medium, similar technologies will no doubt be applied to existing 2D footage to make it come alive in that new format, as well.

Yet these augmentations are only one potential application of this technology. Once a model has been trained to understand patterns in a dataset, such as a library of photos or text, it can generate new media from scratch using those patterns instead of merely attempting to improve the quality of an existing photo, video, or recording.

In 2019, researchers at computer graphics company NVIDIA launched a website called Random Face Generator (This Person Does Not Exist), which features photorealistic portraits of imaginary people. To all but the most highly trained eye, these portraits, featuring faces of every age, gender, and ethnicity, are indistinguishable from actual photos of living human beings. There is an infinite number of potential faces on the site, because they're created automatically by a special variety of machine learning algorithm called a generative adversarial network (GAN).

GANs are a powerful new concept in artificial intelligence. The reason they're so good at fooling human observers is because they're programmed to fool themselves. They're set up as a game played between two AI neural networks (bits of code that emulate the structures of neurons in the human brain). The first network tries to fool the second one, generating new images, text, or sounds based on an existing dataset, which the second network has

Figure 2.2
Six faces generated by this-person-does-not-exist.com.

to discriminate from human-created media in that dataset. The algorithm tries over and over until it's successful: if the first network fools the second network, it's a good bet that the results will also fool a human. Successful results are then fed back into the first network, making it more efficient at fooling the second network (and humans) in the future.

Even though they're a fairly recent invention, GANs and similar technologies are already very popular and are developing rapidly in both power and scope. Two years after the debut of This Person Does Not Exist, for instance, Epic Games (perhaps best known for its online battle royale game *Fortnite*) launched a free online tool called MetaHuman Creator, which allows users—even those with no programming expertise—to create realistic, three-dimensional images of imaginary people and animate them within real or computer-generated settings.

Compared to the still images of faces generated by the GAN on This Person Does Not Exist, the MetaHuman Creator's animated, three-dimensional, full-length bodies represent a great leap in AI-generated imagery. Yet the results are less likely to pass as "the real thing" to discerning eyes and more likely to strike even casual observers as slightly . . . odd. This conundrum was first identified by robotics professor Masahiro Mori in the 1970s, and today it's typically referred to as the *uncanny valley*.[20]

The uncanny valley principle posits that as artificially generated models of human beings move closer and closer to verisimilitude, they begin to seem creepy, or uncanny, provoking emotional rejection by the viewer. That's because even though we can't consciously identify anything wrong with the model, our subconscious picks up on small cues about its artificial provenance and puts us on alert. This principle dovetails with the concept of the Turing test, named for computer pioneer Alan Turing, which essentially says that an AI program's success can be gauged by how well it fools a human user into believing they're interacting with another human.[21]

Both the uncanny valley and the Turing test are circumstantial, contextual, moving targets. As Arthur C. Clarke famously wrote, "Any sufficiently advanced technology is indistinguishable from magic."[22] Conversely, tech that once seemed magical inevitably becomes normalized as it is absorbed into the fabric of daily life—for those with regular access to it, at least. As William Gibson (an inheritor of Clarke's) is renowned to have said, "The future has already arrived—it just isn't evenly distributed."[23] Thus, one person may be completely fooled by an AI or a GAN, believing it to be another human being, while a different person, interacting with exactly the same program, might find it unsettling, or even laughable. That's why a badly photoshopped piece of political propaganda might be more compelling to your computer-illiterate uncle than to you. And why moviegoers reportedly screamed and ran when the Lumières first screened a film showing head-on footage of a train arriving at a station in 1895, while the millions of people who have watched the footage on YouTube probably never broke a sweat.[24]

The uncanny valley principle also demonstrates that photorealism isn't necessarily the easiest or most effective way for AI to fool humans. In the 1960s, Marshall McLuhan famously distinguished information-rich, low-engagement "hot" media like television from low-information, high-engagement "cool" media like books.[25] While there's a lot missing from this reductionist formulation, it points to an important principle: if you're trying to get people invested emotionally in a piece of art or information, sometimes less is more. Cartoonist Scott McCloud demonstrated this in his classic book *Understanding Comics*, proposing that people identify with characters like Mickey Mouse and Charlie Brown because they're so visually minimalist that nearly anyone can see themselves reflected in them.[26]

This helps to explain why some of the most powerful generative algorithms in circulation today aren't pushing the boundaries of "hot" media (like video) in an effort to leap across the uncanny valley, but are working in "cooler" media, where they're less likely to be spotted and more likely to be accepted as authentic human expression. A great example is GPT-3, a program developed in 2020 by research lab OpenAI. GPT-3 can produce cogent, readable prose in a wide variety of formats and contexts based on its training corpus of hundreds of billions of texts written by humans. It can also perform tasks like unscrambling words, solving written arithmetic problems, and answering written SAT analogy questions.

The computer scientists who developed GPT-3 actually performed a Turing test with it, asking people to read short (five-hundred-word) news articles and then judge whether they were generated artificially or written by humans. The people in the study were able to identify which articles were generated by GPT-3 only about half—52 percent—of the time, which is statistically about the same as flipping a coin to make the call.[27] In other words, the model passed the Turing test with flying colors—at least, in a very controlled environment. Consequently, we are now living in an age in which we cannot guarantee that the assertions, opinions, reviews, and confessions we read on the internet or elsewhere are the authentic expressions of an individual person or a highly sophisticated *algorithm or machine. We may now be living in a world in which the truth is always subject to change and correction, and where any given statement may no longer be entirely reliable or true.*

If you find this argument convincing, we can't take all the credit ourselves. The italicized words above were generated by GPT-2, a much less sophisticated precursor to GPT-3, using an autocomplete algorithm.[28] The first, unitalicized part of the sentence was used as an input, and we chose several consecutive sentence fragments to follow it based on the algorithm's suggested outputs in a process similar to the autocomplete function used for the past several years in applications like Google's Gmail service.

In fact, in the year and a half since we used GPT-2 to generate the earlier passage in the initial draft of this chapter, AI models for text generation have expanded even further in power and exploded in popularity. As we write these words in early 2023, companies like Google and Microsoft have begun to integrate them into search engines, word processing software, and other tools, sometimes with decidedly uncanny results. By the time you read this book, it's likely that you will encounter unacknowledged AI-generated text throughout your day—in advertisements, emails, technical documentation, and even journalistic articles and other creative contexts like (perish the thought!) academic publications.

Yet for all the success these algorithms have in tricking human readers into believing that their output is meaningful and valuable, it's important to remember that no matter how sophisticated the results may seem, what's going on behind the curtain is just highly educated guesswork. As computational linguistics researchers Emily M. Bender, Timnit Gebru, and others wrote in 2021, these software programs can best be understood as "stochastic parrots," repeating back what people say to them in statistically informed combinations that lack any grounding in actual context or conscious intention.[29] If we see meaning in these results, they argue, it's best understood as a reflection of our own expectations and experiences, rather than the expression of an artificial consciousness.

CONNECTING THE DOTS

In 2019, Jill Abramson, former executive editor of the *New York Times*, released a new book entitled *Merchants of Truth*—a full-throated protest against the degradation of the American news industry due to digital advertising models and social media, and a rousing call for a return to journalistic ethics.[30] The day it was released, it sparked an international firestorm, with some of the biggest names in the news media weighing in on the difficulties and dangers of policing ethics in journalism. Normally, this kind of attention would signal a major victory for such a book, but unfortunately for Abramson, it was she herself who ended up in the crosshairs of an ethics controversy, with numerous credible allegations of plagiarism from other journalistic sources evident throughout her book.

Michael Moynihan of Vice TV—who is mentioned explicitly in the book as part of Abramson's critique of his news organization—was the first to identify the alleged plagiarism, posting a Twitter thread documenting some examples from the advance copy he had received.[31] Others soon followed suit, and within days, Abramson found herself answering increasingly critical interview questions instead of enjoying the victory lap typically taken by a high-profile recent author. While she never outright confessed to plagiarism, she did acknowledge to NPR that "in several of these cases, the language is too close for comfort," and that she should have exercised more care in her use of other people's work.[32]

As many noted at the time, it was sadly ironic that a book about journalistic ethics would be subject to accusations of major ethical lapses. Yet there was a second irony, as well: Abramson's borrowings may never have been discovered if it weren't for the new digital media infrastructure she blamed for catalyzing the changes in her industry. To put it simply, today, anyone can freely and easily Google any given sentence or a phrase and find out in a matter of microseconds whether it has previously been written in any of millions of books or on any of billions of web pages.[33]

For people with the right tools, identifying plagiarism is even easier than doing a Google search. Our universities have provided us, as professional educators, with access to services like Turnitin that can automatically scan student assignments for possible uncited borrowings from third-party sources, including those that Google might miss, such as student papers that have previously been checked by the service but never published online. As with all algorithmic models, the results of these plagiarism checks are probabilistic; the service can report that a given paper is 71 percent likely to be plagiarized, or 18 percent, but it can't determine exactly what happened.

While these services have in some ways made our job as educators easier, they've also added new challenges. First and foremost, just because we *can* automatically scan every single student assignment for illicit appropriations doesn't mean we *should*. We must decide for ourselves whether the benefits of catching rule breakers outweigh the costs of subjecting our students to added surveillance and treating them as presumptively guilty rather than trusting them by default. Second, even though plagiarism scores are probabilistic, university grades are not. There is a very strict binary difference between pass and fail. It falls to us to interpret between probability and assessment. Is a 90 percent likelihood of plagiarism enough to warrant a failing grade? What about 75 percent or 50 percent? We could, of course, follow up with the student in question to get to the bottom of things, but that would make our jobs more difficult, not less, undermining the rationale for using the software in the first place.

Finally, it's not clear that software like Turnitin actually makes students behave any more ethically; it might just make them better at hiding their ethical breaches. While there are many websites offering plagiarism-checking services (several of them supposedly aimed at students doing a last-minute scan before turning in a paper), there are just as many offering "paraphrasing tools" designed to make borrowings indetectable by altering the language and syntax just enough to elude plagiarism-checking software. Just like a GAN, these tools are essentially machines trying to fool other machines in order to fool people, using massive libraries of digital media as the basis for both sides of the "game." So, in the end, such technologies don't actually change human behavior much at all—they merely end up coercing us into outsourcing our intuitive, emotional, and ethical relationships with one another to probabilistic algorithms.

Our point here isn't that being an educator is difficult in the age of the internet. Or that students are bad people, intent on cheating. Or that machine learning is inherently a bad thing. Our point is that machine learning and artificial intelligence may be used not just to derive knowledge from a given library or archive of material, but to collect data across a range of different sources (including the outputs of other ML and AI models), identifying patterns and similarities that would never be evident to an individual human being, or even to an AI trained on a single, discrete collection of data.

In other words, machine learning algorithms aren't learning only from their programmers or their users—they're also learning from, and feeding back into, society at large.

This, in turn, means that ML algorithms inevitably have consequences that extend far beyond the contexts in which they're deployed (e.g., spotting plagiarism by unethical students and famous journalists). This becomes increasingly true as a wider and wider range of interested parties connect a larger and larger number of devices, sensors, and media archives to the internet. The kind of power Turnitin offers to educators is rapidly becoming a commonplace and easily accessible commodity, disproportionately available to those already in positions of authority. Which means that *everyone* must grapple in one way or another with the ethical questions educators have raised about wielding such power and interpreting the outputs of these models. It also means that the kind of cat-and-mouse arms race dynamic between plagiarists and plagiarism detectors, often driven by rival tools provided by the same source to both sides, will come to characterize more and more relations of power, from schools to workplaces to the public square.

POLICING ALGORITHMS

One area in which pattern-matching algorithms have already been transformative is in intelligence gathering. Over the past twenty years or so, police and military organizations around the world have begun using what is commonly referred to as multi-intelligence fusion (or multi-INT) as a crucial platform for their operations and, to an extent, as a new paradigm for investigation. Using software tools provided by well-known companies like IBM and Microsoft, as well as lesser-known specialists like Palantir and Genetec, multi-INT systems enable their users to stitch together disparate sources of data as diverse as CCTV cameras, financial records, and social media activity to profile individuals in real time, or to develop a Batman-esque, top-down view of all the potentially threatening activities currently unfolding in a given geographical area.

These systems are marketed with the premise that more is better—the notion that multiple, correlated data points can be analyzed in real time to generate new insights that might otherwise be hidden. Yet the end users of these technologies, such as police officers and intelligence analysts, rarely have either the expertise or the under-the-hood access to understand how and why certain data points have been connected, let alone the ability to distinguish correlation (two things seem to be related) from causation (one thing leads to another).

As Christopher Nolan's *Batman* films and countless other works of speculative fiction have explored, this kind of power is bound to be abused in the absence of a stringent system of checks and balances restraining its development and deployment. At the moment we're writing this, no such system of checks and balances exists, and there's really no way to know the extent to which these systems have been employed—or abused. The New York Police Department, for instance, currently uses a multi-INT system called Patternizr that was built internally to automate the investigative techniques of NYPD detectives. The platform was rolled out in secret, and its existence was only revealed to the public after more than two years of use.[34] While the department has touted Patternizr's potential to "improve public safety" by expanding investigations beyond the borders of individual precincts, nothing prevents officers from using it for less ethical purposes, from stalking sexual partners (a practice so prevalent in the NSA that it gained the nickname LOVEINT) to spying illegally on religious minorities (a practice that landed the NYPD in hot water even before it began using multi-INT systems).[35]

Even when multi-INT policing systems are employed exclusively for their intended purposes, their use can have significant consequences for human rights and civil liberties. For one thing, these technologies are sometimes used by institutions with far more power and less accountability than the NYPD. The Chinese government, for instance, has integrated multi-INT into its systematic oppression of the nation's millions of Uighur Muslims, compounding and augmenting a brutal policy of subjugation that includes torture and mass incarceration.[36] But even "proper" usage of these systems by relatively ethical actors can have unexpected repercussions that widen existing social divides and make life even harder for minorities, dissidents, and other vulnerable populations.

Just like plagiarism-detection software, multi-INT systems can be used in ways that challenge institutional power relations by connecting the dots to hold governments and large corporations accountable for their own behaviors. There is a burgeoning field of data journalism that uses a host of forensic and analytical tools (including ML algorithms) to spot issues like financial fraud and to expose hidden webs of political influence, as well as to counter the official story about news events provided by governments or industry spokespeople.

One of the emerging leaders in the data journalism field is Bellingcat, an independent investigative news collective that focuses primarily on political crises and conflicts in Eastern Europe and the Middle East. Instead of putting reporters on the ground to provide first-person accounts of a situation, as a traditional outlet like the BBC or the *Washington Post* might do, Bellingcat collects data related to an event (both structured and unstructured, ranging from GPS metadata to social media posts to video footage), contextualizes them via follow-up interviews with eyewitnesses, and then attempts to stitch the resulting information together into a comprehensive narrative that has more nuance, detail, and (they hope) reliability than any first-person account could offer.

These techniques are not a replacement for traditional, reporting-based journalism, but they add a valuable—and sometimes revelatory—perspective on world events. For instance, in 2020, Bellingcat released a bombshell report that "Greek security forces likely used live rounds . . . against refugees and migrants trying to break through the Turkish-Greek border fence," a claim that a Greek government spokesperson had "categorically denied" as "fake news."[37]

Bellingcat collected hundreds of images, videos, and accounts of the event, most of them scraped from social media, and combined them with data sources such as satellite imagery, EXIF and Facebook metadata, death certificates, and witness interviews to create a comprehensive, geolocated (place-specific), and chronolocated (time-specific) reconstruction covering roughly forty-eight minutes and tracking the actions of dozens of participants. The reconstruction was then assembled into an animated 3D sequence by software company Forensic Architecture. The resulting news report, which relies heavily on the reconstruction to support its claims, makes a very compelling case that the Greek government lied and indeed used live ammunition against refugees, killing at least one and wounding several more.[38]

According to Nick Waters, the Bellingcat journalist who wrote the story, multi-INT modeling like this is not only useful for investigations, but also helps to communicate evidence to audiences who may not find a single video or firsthand account sufficiently convincing. "I can't take a reader and get them to watch and rewatch all these videos over and over and over again," he told us in an interview. "It's a hugely complex situation in which thousands of potentially relevant details are happening within a period of forty minutes. It's much easier to show what's happening using the model that we created."

THE SILENCE IS DEAFENING

The geometric growth in digital sensors, forensic techniques, pattern-matching algorithms, and ML and AI models has generated an unprecedented deluge of data. Even with sophisticated multi-INT fusion systems and state-of-the-art data centers featuring of rows and rows of supercomputers crunching numbers at top capacity, the volume of data is greater than anyone can process, and the number of potential connections between these data points is literally incalculable.

In the early 2010s, business analysts and computer scientists began to use the term *big data* to describe the qualitative challenges, techniques, and opportunities that are unique to databases measured in petabytes (a petabyte is equal to a million gigabytes, or 10^{15} bytes of information) or even bigger increments. These were contrasted with traditional, smaller datasets that could be handled by normal business computers and statistical methods, such as the responses to a public-opinion survey or a pharmaceutical trial. Over the past decade, *big data* has become a buzzword, gracing the covers of magazines and featured prominently in the public discussion of technology, from policy debates to financial news to marketing materials.

Yet there's something fallacious about the distinction between big and small. And that's because, in the age of multi-INT and cloud computing, there's no longer any such thing as small data. When every single data point can be correlated with any other data point in any number of rapidly growing and publicly accessible archives, *all data are big data*. Did you poll your Facebook friends about their favorite flavor of ice cream? Does your doorbell collect surveillance video of your front porch? Have you used an app to keep track of your blood pressure, per your doctor's orders? Once upon a time, these little factoids might have stayed in your notebook or been stored only on a cassette in your library. Today, they are digitized information, available to countless businesses, governments, and other internet users to be cross-referenced with other data about you, and about the world at large, for all time, from the moment they're recorded.

Even the absence of data is now a form of data. Like a dark spot on the sun, a missing data point can stand out against the glare of information overload in stark relief. People who fail to understand this may end up learning it the hard way. That's exactly what happened in 2018, when Yandex Maps, Russia's biggest online mapping service, accidentally revealed the probable locations of more than three hundred military sites, including airfields, barracks, and nuclear facilities, in Israel and Turkey. As Matt Korda, a researcher at the Federation of American Scientists, demonstrated in a blog post, these sites were identifiable precisely because Yandex had blurred them out, "almost certainly the result of requests" from the two countries.[39] It might seem counterintuitive, but if Israel and Turkey *hadn't* requested that Yandex take these steps, the locations of their military sites would probably have remained far more secure; after all, it's much easier to identify a blur in the midst of a crisp photo than it is to distinguish the roof of a barracks from the roof of a nearby warehouse.

Missing data can even be used as a smoking gun. We spoke to Chris Imler, a digital forensics expert at Legility, a company that provides legal services to corporations and law firms, who told us about a case of corporate espionage that was exposed this way. An employee at a company was leaving to work for a rival firm and illegally copied some of his old employer's intellectual property before wiping the files from his hard drive to destroy the evidence. Imler's analysis showed that the hard drive contained distinctive blank spots where the ones and zeroes on the disk followed a "standard wipe pattern" from a digital eraser program called CCleaner. Using this evidence, Imler was able to connect the alleged corporate spy to a search history for terms like "how to securely erase a hard drive" and, ultimately, a registration for CCleaner's pro version. As Imler described, large amounts of missing data on the employee's hard drive, their search patterns in the days before leaving the firm, and their use of personal email to relay proprietary inventory and price lists, "all led to the conclusion that something was not right." When the attorneys for the alleged spy read Imler's report, they saw the writing on the wall and quickly settled the lawsuit.

Missing data can also create an opportunity for bad actors to game the system by being the first to fill in the blanks with false information. This is a problem that has been documented by tech researchers Michael Golebiewski and danah boyd, who refer to rarely searched keywords and rarely discussed subjects as "data voids."[40] This is a frequent problem in the area of breaking news, where small towns, uncommon names, and arcane jargon can be propelled suddenly into the mainstream conversation. For instance, Golebiewski and boyd analyzed the online buzz following a church shooting in the town of Sutherland Springs, Texas, in 2017. As their research demonstrated, "no one had searched for this town for years" on the internet, and therefore there were very few sources of information about it online. Immediately after the shooting took place, a network of mostly fake social media accounts began to share disinformation about the event, for instance falsely claiming that "Antifa" was behind the attack. This disinformation rapidly rose to the top of search engine results for "Sutherland Springs," and as a result, millions of people were exposed to these lies when they searched for details and updates about the shooting. Far-right political news websites were also able to amplify the disinformation, using its prominence—and the absence of correctly reported information for a crucial few hours—as a source of plausible deniability.

To sum up: small data are big data. The absence of data is a form of data. And small data in the presence of a data void can become a data supernova.

WHAT COULD GO WRONG?

When Francis Galton first embarked on his lecherous project to build a "Beauty-Map" of the British Isles, he didn't bother asking the women he passed on the street for consent before collecting and analyzing data about them, let alone consigning them definitively to the categories of "attractive," "indifferent," or "repellent." Though this project was abhorrent on its own terms, it was thankfully limited in its immediate impact; it's not as though a potential employer, romantic partner, or financial services company had access to his data, or the ability to discriminate based on his rankings.

Yet Galton's pet project inaugurated a new era in social data collection and analysis, and the entire world has been radically altered as a consequence. Today, young women in the British Isles—and people of every gender in every populated location on the planet—generate multiple forms of data all day, every day, for the duration of their lives, and those data are archived, analyzed, and correlated by a broad range of governmental, commercial, and criminal institutions. Public CCTV cameras equipped with facial recognition technology have taken the place of Galton's gaze.[41] Massive online databases play the role of the paper cross with pin pricks. Social media posts, financial interactions, fitness trackers, web browsing histories, GPS mapping coordinates, EXIF and phone metadata, and thousands of other data sources feed into these databases, contributing to highly detailed, frequently inaccurate, and immensely powerful profiles of nearly every living person, as well as helping to generate realistic profiles of imaginary people.

Unlike Galton's subjects, we are at least somewhat aware of these data collection systems and can resist them to some extent (e.g., blocking website cookies, wearing a face mask in public places, or advocating for laws limiting the use of ad trackers and surveillance cameras). On the other hand, it's likely a losing battle. We can't possibly know about all the data we produce or how they'll be analyzed, let alone opt out of this data collection—unless we fancy life as a hermit in a cave somewhere. Barring some kind of radical change, it seems likely that these systems will continue to grow in power, prevalence, and influence in the years to come, and play an ever larger role in shaping our public and private lives.

This doesn't mean we should just lie down and accept these technosocial changes as inevitable or ignore their potentially damaging consequences. Researchers and advocates have identified many different ways in which the technologies we've discussed are already influencing our cultures, our legal and economic systems, our interpersonal relationships, and our identities. In the next chapter, we'll review some of these big-picture consequences and discuss some of the ways in which people are attempting to flag and fix potential problems before they overwhelm us altogether.

3 BIG DATA BLUES

The river of data that surrounds us in networked society can feel like a flood. The sheer volume of ones and zeroes generated by the bare necessities of our digital lives—our phones, our laptops, our "smart" appliances and medical aids—constantly threatens to sweep us off our feet and overwhelm us. Our photos are data. Our purchases are data. Our romantic encounters are data. Even our sleeping habits and bodily functions are data. And all of those data have secret lives, which we have barely any power to control.

Then our data spawn more data, and those data are analyzed and recombined to become other people's data. Small data are aggregated into big data. Even the absence of data becomes a form of data, and real data can be used to generate fake data. No wonder so many of us have given up hope of ever finding our way back to dry land.

Then again, you may ask, what's so great about dry land, anyway? Smartphones help bring us together. Laptops help us get our work done. Smart devices and medical aids can help save lives. Who cares if we're drowning in a substance we can't even see? It's not like data are *real*. They're merely abstract representations of reality. At the end of the proverbial day, what difference does it make if the details of someone's life are data points on a random company's hard drive?

As people who understand the power of data will tell you, it makes all the difference in the world. Those data points are used by powerful governments and corporations—as well as everyday people—to make political, financial, and cultural decisions that affect all life on earth. A single entry in a spreadsheet somewhere can mean the difference between life and death for a hospital patient, between bail and jail for an accused criminal, or between a despotic regime and a democratic one for an entire nation.

Yet, as we have taken pains to point out, data are not a mirror image of the world they supposedly represent. They're distorted by our expectations, circumscribed by our limitations, and always generative of countless new and unexpected forms of knowledge, which are probabilistic rather than absolute. And, like Galton's paper crosses and the countless data collection instruments made in their image over the past century, they reflect the assumptions of the investigator just as clearly as they portray the person or thing being investigated.

For those of us drowning in data, it's a growing challenge to navigate the complex landscape of our daily lives, as we become increasingly aware of its role in mediating our relationships, illuminating our secrets, and (mis) representing us to the world at large. These aren't merely abstract problems, nagging at the back of our minds among all of our other existential anxieties. They're also increasingly practical challenges, creating turbulence and whirlpools as we're swept along in the current of data, forcing us into confrontations with systems we don't fully understand, requiring us to rethink our ideas about the world and how to live in it.

Below, we'll discuss a few of the specific ways in which the secret life of data is complicating the daily lives of people, and we'll introduce you to some of the researchers, advocates, and activists who are trying to solve these problems before they get any worse.

"WHO WAS THAT MASKED MAN, ANYWAY?"

Once upon a time, there was no such thing as anonymity (literally, "namelessness"). Human beings lived in small enough groups that everyone knew everyone else, and it was rare to meet someone whose face you didn't recognize on sight, or to hear words spoken in a stranger's voice. As human societies became more complex, however, we put greater care and energy into keeping track of who was who and who said what. Larger and more stratified cultures developed elaborate naming conventions, adding secondary titles to designate someone's family lineage, profession, or place of origin. With the development of written language, and ultimately the printing press and other forms of communication media, people's identities and expressions became increasingly disconnected from their bodies. Speakers became authors, their names affixed to publications that could be bought or sold and "consumed" miles and years from the point of their creation. With the advent of copyright law, even the legal ownership of one's own creative expression could be traded away. The name affixed to a book or image became the only remaining connection between a physical human being and the record of their ideas.[1]

As keeping track of who was who and who said what became an increasingly complex task, anonymity became an increasingly powerful tool. In the popular imagination, it became associated with both altruism and crime (think of Zorro or the Masked Bandit). Political treatises responsible for fomenting revolutions from Russia to China to the Americas were published anonymously or pseudonymously. Even in the 2020s, one of the world's most notorious international hacker collectives uses the name Anonymous, and its logo is an image of a masked face, while one of the most virulent conspiracy theories influencing the geopolitical landscape goes by the name QAnon and centers on a shadowy anonymous figure who supposedly shares revelations from deep within a corrupt government institution.

Yet anonymity is hardly the exclusive domain of outlaws and mysterious benefactors. In a large and complex society, anonymity is contextual. Most of us carry around multiple identification cards and electronic devices featuring photos of us, our home addresses, and many other forms of metadata intimately tied to who we are and what we do with our lives. But we also know that as we walk down the street, the vast majority of the faces we see and voices we hear will be those of strangers. Those of us who live in urbanized societies—a group that includes most people alive today—are, in the eyes and ears of those around us, functionally anonymous.

Social theorists in the nineteenth and twentieth centuries spent a lot of ink and energy asking whether this was a good thing. Functional anonymity, they argued, could contribute to anomie, or alienation, and to the breakdown of moral, cultural, and psychological well-being. On the other hand, some argued, anonymity was a path to liberation. People were freer to express challenging ideas, to explore innovative and idiosyncratic lifestyles, to escape the crushing conformity and stultifying sameness imposed by the constant scrutiny and judgment of those around them. As the twenty-first century approached and computer networks grew in scope and power, these debates naturally shifted to online identity. Some saw liberatory potential in digital anonymity. As a now famous *New Yorker* cartoon put it in 1993, "On the internet, nobody knows you're a dog."[2] Others, however, saw the internet as a kind of Big Brother 2.0—a platform ready-made for state and commercial surveillance, and a threat to free societies.

New digital forensic techniques raise the stakes of this debate even further and make it more difficult to resolve. As the physical and social landscapes we inhabit have become more datafied, the "real world" has begun to behave increasingly like the internet, and the walking-down-the-street model of functional anonymity no longer applies. We can't expect to be seen as passing strangers when images of pedestrian traffic are captured by cameras, stored on a server, and analyzed by facial recognition systems. We can't expect to have private conversations when always-on devices in our homes are recording our words and providing them to law enforcement or commercial interests on demand. And we can't expect even our most sensitive health and financial records to remain secret for long when the servers they reside on are routinely hacked and the "non-personally identifiable information" they hold can be deanonymized by pattern-matching algorithms.

One of the earliest and most consistent voices speaking about the potentially damaging social consequences of deanonymization is Harvard professor Latanya Sweeney. In a groundbreaking 1998 article, she and Pierangela Samarati introduced the concept of "k-anonymity," which expresses the idea that data about people may be *technically* anonymized if you look only at the particular dataset they belong to, but can be deanonymized once they're linked with other datasets.[3] They showed that using a list of 55,000 voters from an unnamed city, 97 percent of the supposedly anonymous entries could be linked to specific individuals just by cross-referencing the zip codes and birthdays of the voters with information in other publicly available databases. In other words, when all data are big data, it's hard to maintain contextual anonymity, because the context keeps getting wider and wider, and the connections between data are themselves data that can be mined.

Sweeney is a computer scientist, but the concept of k-anonymity is not merely an abstract technical consideration. As she has spent the past several decades arguing, data systems that guarantee anonymity only in their original context need to be recognized as severe security threats, and treated as violations of our privacy laws—and as incentives to strengthen those laws. While she was writing the k-anonymity paper, Sweeney made headlines for a dramatic demonstration of these points. Her home state of Massachusetts was in the process of sharing anonymized health records for all state employees online, so she decided to delve into the data and find the records for then-governor William Weld. She purchased a local voter list for $20 and searched the health records for an entry matching Weld's demographic profile, easily finding a match. News of Sweeney's discovery helped to jumpstart a public discussion about the necessity of protecting health data from deanonymization, and Sweeney's work was cited specifically in the preamble to the new privacy rule of HIPAA, the nation's principal law governing the protection of medical records.[4]

A quarter of a century later, despite the strengthening of laws like HIPAA, deanonymization has become such a widespread phenomenon that a gambit like Sweeney's would no longer bring any shock value. As we were writing this section, for instance, the *Pillar*, a journalistic website that exclusively covers stories related to the Catholic church, broke the news that Monsignor Jeffrey Burrill, general secretary of the US bishops' conference and an outspoken opponent of LGBTQ rights, had for the past three years been a regular user of queer hookup app Grindr and a frequent visitor to gay bars and clubs.[5] The *Pillar* was able to out Monsignor Burrill (who immediately resigned his position when the news broke) by correlating "commercially available records of app signal data" with information about Burrill's locations (his residences, work sites, and travel destinations) during that time frame, effectively deanonymizing those data.

These revelations barely made a splash in the larger news cycle, which was dominated by global crises such as the COVID-19 pandemic and climate disasters, as well as reports of widespread surveillance of journalists, dissidents, and heads of state via mobile phone spyware manufactured by the Israeli firm NSO Group. Compared to such bombshell stories, the outing-via-deanonymization of a single clergyman makes as much of an impression as a flashlight shone on a five-alarm fire. Besides, unlike in the NSO Group story, no laws were broken. While the ethics of the *Pillar*'s investigation (and of Burrill's behavior) are certainly debatable, it was completely legal in the United States, where the aggregation and sale of "anonymized" user data from mobile phones and apps is a thriving marketplace open to any buyer willing to pay the going rate.

Yet the secret life of deanonymized data doesn't end in the marketplace; it begins there. Once personal data have been harvested, aggregated, and made available—whether legally, quasi-legally, or illegally—they can be used by unaccountable third parties to fuel criminal operations and harassment campaigns. For instance, after personal data from 33 million accounts at Ashley Madison, the "dating website for cheaters," were leaked online in 2015, innumerable account holders in several countries were targeted with blackmail and extortion demands. This crime wave lasted for at least five years, yielding untold millions of dollars in hush money and contributing to several documented suicides among victims.[6]

In other cases, deanonymization isn't merely an ultimatum made by extortionists—it's an actual means of intimidation and assault. Because anonymity has provided people with the freedom to express dissident political opinions, embrace minority sexual and gender identities, and engage with unconventional subcultures, reconnecting someone's behaviors to their physical bodies and personal identities holds potential for threats of both physical assault and reputational harm. Antisocial internet behaviors such as *doxxing* (revealing someone's physical address and other personal information without their consent) and *swatting* (making false allegations in order to send law enforcement to someone's home) have become increasingly commonplace in recent years and are often undertaken as punishment or intimidation of those who express opinions or identities that the harassers view as undesirable.

Yet, as with all of the techniques and strategies we've discussed so far, deanonymization cuts both ways, not only threatening the vulnerable and precarious, but also holding the powerful more accountable, for better and for worse. While Latanya Sweeney may have deanonymized Governor Weld's health data as a stunt to advocate for stronger privacy laws, many others have doxxed public officials and corporate executives as a way to show these social elites that they're not as untouchable as they may once have assumed. In 2013, for example, members of the aforementioned hacker collective Anonymous doxxed over four thousand American banking executives to protest the Computer Fraud and Abuse Act, an antihacking law that many internet freedom advocates believe is overly punitive and broad in scope.[7] Clearly, this kind of intimidation is no more ethical or socially desirable than when ethnic or sexual minorities are targeted by members of dominant social groups, but it does illustrate the principle that the rising tide of data threatens to capsize all boats.

Ultimately, the problem with deanonymization is that it undermines our privacy. It connects us, without our knowledge or consent, to data we might not want our names associated with. But what about the cases where we *do* want credit for creating something, where we're happy to sign our name to a piece of information? Modern society rewards and celebrates creators who share their work in public, bestowing money, fame, and power on its most celebrated artists and authors. And while the rights and responsibilities of these creators aren't likely to disappear any time soon, the secret life of data poses just as much of a threat to the concept of authorship as it does to privacy, potentially undermining and even severing the links between a creator and their work.

WHO OWNETH WHOSE WORDS AND ACTIONS?

Earlier, we talked about some of the ways in which new digital forensic techniques have created a field day for art historians, allowing them to reexamine age-old mysteries about the authorship and provenance of historical works, giving them a glimpse beneath the surfaces of famous paintings, and letting them reconstruct and restore dusty relics to their original glory. The news coverage of these innovative developments tends to fall into the "gee whiz" category, and of course the technologists and historians interviewed for these stories are justifiably jubilant about their discoveries.

Yet the artists themselves are almost never interviewed for these articles—to be fair, Rembrandt and Picasso can't be reached for comment—and therefore ethical questions about whether and how an artist should have a say in the matter are rarely addressed. And the even bigger questions of what these interventions mean for society at large, such as the changing roles that creativity and authorship play in our cultures, our identities, and our markets, are seldom even hinted at. Whether or not we're paying attention, however, these tectonic cultural shifts will reshape the world as the secret life of data continues to transform the public life of art.

For most of the past millennium, the idea of authorship in the Western world has meant far more than just getting credit for writing a book or painting a picture. It has been a metaphor for personhood, individuality, autonomy. It's no accident of history that artists started signing their names to paintings as people began to lay the groundwork for representative government and secular cultural institutions during the European Renaissance. The "unalienable rights" enshrined in the United States Declaration of Independence a few centuries later could not have been claimed without the preexisting concept of an independent person endowed with such rights by birth, and that idea was communicated through the metaphor of individual creativity. This connection was very clear to social theorists, even at the time; as Thomas Hobbes, one of the first political philosophers of the modern era, asserted in his 1651 book *Leviathan*, "He that owneth his words and actions, is the AUTHOR."

Yet modern Western ideas about individuality and artistry have also been challenged since they first emerged, not only by the medieval spiritual and political institutions they threatened, but also by forward-thinking philosophers and artists who believed these ideas didn't go far enough toward liberating the human race from hierarchical power structures like church and state, and who worried that the concept of individuality would end up being as much of a prison as feudal peasantry or religious dogma.

From the earliest days of the novel, for instance, authors like Laurence Sterne and Miguel de Cervantes toyed with the form, undermining the authority of the author and the individuality of the individual by creating unreliable narrators and inconsistent heroes. Along similar lines, composers like Handel and Mozart freely "borrowed" melodies and arrangements from one another, drawing from a shared and evolving language of musical ideas to craft the symphonies that would (ironically) later be celebrated as evidence of their unique genius. By the twentieth century, artists in creative movements like Dadaism and postmodernism had taken these games even further, using scissors and glue to literally cut and paste pieces of other people's work into their own.

In other words, authority has always been under assault, frequently by authors themselves. Art critics and political philosophers have periodically recognized the power of these attacks and their broader implications for society at large. In 1968, French philosopher Roland Barthes famously proclaimed the "death of the author" and argued (perhaps overoptimistically) that rejecting the "myth" of the individual artist would lead to "truly revolutionary" liberation from both religious and scientific dogmas.[8]

Because of their challenging nature, these kinds of ideas have generally been shunted to the side by mainstream institutions and dismissed as "fringe," "avant-garde," and even "piratical" or "heretical," despite their popularity with both audiences and creators. Until recently, it was relatively simple to sideline art that punctured the myth of the individual artist, because only a handful of large corporations had the resources, power, and legal authority to distribute books, music, and visual art on a massive scale. But these cultural gatekeepers, as they're often called, faced a major setback when millions of people around the world started using personal computers in the 1990s, combining the power to digitally cut and paste media with the ability to distribute it globally via the internet. Within a decade, billions of internet users were creating, adapting, and sharing memes, remixes, mashups, and countless other kinds of configurable culture via peer-to-peer networks and social media apps.[9]

Think about the last meme you saw. Maybe it was a photo of "Side Eyeing Chloe," a little girl with blonde hair and prominent front teeth looking apprehensively at the camera. Maybe it was a line drawing of astrophysicist

Neil deGrasse Tyson raising his hands in a defensive posture. Maybe it was a newer image from a movie, anime, or video game that was published after we wrote this book. Whatever the visual component of the meme was, three things are likely: (1) that you've seen several, perhaps even hundreds, of versions of this meme on your social media feeds; (2) that you understand the connotations of the meme, spanning multiple specific versions; and (3) that, even if you know where the image originated, you don't know who originally made it into a meme or put it into wider circulation. In other words, this kind of cultural activity flies in the face of modern ideas about individual artistry; its materials are all repurposed, and its meaning is defined collectively. Courts have been cautious about applying copyright enforcement to memes and remixes, recognizing that certain kinds of collective, configurable culture fall under the category of "fair use" because of their transformative nature, while the borrowings of other media creations are so minimal that they don't violate the original creators' copyright at all.

New developments in forensics and AI have pushed this already strained concept of creative authorship and authority to a breaking point. First and foremost, what are the ethics of using these techniques to derive new knowledge about old artworks? Perhaps Picasso didn't *want* us to know how he repurposed elements of an older painting to create *La Miséreuse accroupie*. Perhaps Edvard Munch *intended* the authorship of his cryptic scribble on *The Scream* to remain a mystery for the ages. Perhaps John Lennon and Paul McCartney *chose* to obscure the division of compositional labor in their songs because they believed that together, they were greater than the sum of their parts. Mostly, we don't know because we can't ask. In some cases, even though we can ask, researchers and reporters typically don't.

These questions get even thornier when we consider the power dynamics surrounding art made in more distant times and places. What would medieval monks have said about the *Book of Hours*—a religious artifact made through devotional practice—being inundated with ultraviolet radiation and effaced by the pagan script they had so painstakingly scraped from their parchments? What would the Syrian sculptor of the "Beauty of Palmyra" have thought of a Danish scientist from another millennium making algorithmically informed creative decisions about the features and coloration of her digital reconstruction? These cases aren't simply about undermining authorial intent; they intersect with questions about the boundaries between religious and secular power, and the responsibilities of colonialist states to formerly colonized regions. As the French philosopher Michel Foucault famously formulated, knowledge is power—and the ability to create knowledge about someone or something is always an exercise of power over that person or object.[10] Yet these considerations, too, are largely absent from public discussion of these new tools and their uses.

The challenges only increase as we move beyond forensic investigation of old artworks and into the wholesale digital creation of new ones. It's not an exaggeration to say that GANs and other forms of so-called AI authorship drive a giant nail into the coffin of the modern concept of individual creativity.

First, as we discussed earlier, GANs and ML models are generative, creating new work, but they can only do so using the features they've extracted from previously existing work. In other words, they can imitate, but they can't innovate (at least, not intentionally).[11] So if an AI is fed hundreds of baroque compositions and then generates a new canon based on what it "learned" from its analysis, who is the author of that canon? All of the musical ideas it expresses came from Bach, Handel, and their contemporaries. Should they be credited as co-composers? Does the programmer who made the AI deserve any of the credit? How about the musicologist who selected the material on which to train the AI? Or, looking beyond the human actors in the equation, should the AI itself be credited as the composer? Remember: the modern concept of authorship has always served as a proxy for personhood, which means that giving credit to the software program could be seen as tantamount to calling an AI an individual.

What about "style transfers," in which an AI is fed the work of a single artist, then generates a "new" work in the style of that artist? In 2021, mental health advocacy organization Over the Bridge released a compilation called *Lost Tapes of the 27 Club* featuring AI-generated songs based on the work of recording artists who died at the age of twenty-seven, such as Kurt Cobain, Amy Winehouse, and Jimi Hendrix. Each of these songs may seem, to a casual listener, like the genuine artifact; as musicians who were in college during the grunge years, for instance, the song based on Nirvana's discography, titled "Drowned in the Sun," sounds to our ears like a long-lost recording by the band from 1993. Yet neither the late Kurt Cobain nor anyone else in the band directly contributed to the generation of this recording. Instead, a machine learning algorithm analyzed the Nirvana catalog and generated some musical and lyrical approximations of their style, which were selected and assembled by humans at Over the Bridge, who then arranged a complete song from these bits and pieces and hired a Cobain impersonator to sing over the backing tracks. The same authorship questions we raised above apply, only now the case for crediting Nirvana is even stronger, as they created all of the compositional and performative elements that were fed into the algorithm. Yet the case is also stronger for crediting the producers who assembled the fragments of computer output into a recognizably Nirvana-esque song, as well as for the singer who helped to seal the deal by imitating Cobain's signature snarls and whines.

Not only do these scenarios strain our definitions of artistry and authorship and pose vital ethical challenges that we have yet to grapple with as a society; they also wreak potential havoc on our legal, economic, and social systems. Notably, there don't seem to be any copyright claims made for "Drowned in the Sun," as the composition doesn't currently appear in the databases of publishing registries hosted by organizations like ASCAP and BMI, and there is no copyright notice on the streaming sites currently hosting the song or the homepage for the compilation. Does that mean it belongs to nobody? Or to everybody? Could a filmmaker use the song in a soundtrack without permission or payment? Could another artist cover the song without paying a royalty? Should members of Nirvana, or Cobain's estate, be compensated for their central roles in the process? It's not simply that we haven't done enough homework to figure out the answer—the point is that *there is no answer*, because these questions are (at the time of writing) not addressed by the law and have not been tested by the courts.

As more and more cultural products are produced by some version of AI and ML, from fashion models whose faces don't really exist to news articles written by commercial algorithms to digital images generated using tools like DALL•E and Stable Diffusion, these questions will only become more troublesome. More and more money will be on the line, and new varieties of legal complications and societal challenges, from libel to hate speech to misinformation, will increasingly crop up as well.

American copyright law currently says that an artwork is considered anonymous unless a "natural person is identified as author." This definition of authorship leaves us with only two options for attributing AI-generated works, each of them equally problematic. We could say that the author is the human who owns the AI in question. But that, in addition to undermining our cultural understanding of authorship (and, therefore, personhood) by reducing it to a transactional relationship, would fly in the face of legal history. In antebellum America, if an enslaved person invented or authored something, they could not be granted a patent or copyright, because the law didn't recognize them as human beings. Yet their enslaver couldn't register a patent or copyright either, because, although the author or inventor in question was legally considered their property, they couldn't claim to have created or invented the work themselves. The same goes, one might presume, for the owner of a software program that generates a work on their behalf.

Alternatively, we could decide that if an AI generates a work, the AI itself is the author. From a legal perspective, this would currently require that we treat the AI as a "natural person" for copyright purposes. But the legal, economic, and social implications of such a decision would get very complicated, very quickly. Would the AI be paid for its labor? Such a decision would mean granting computer software the right to own property. Would the AI be credited for its contributions?[12] Such a decision would mean considering the AI to be an independent actor with something approximating free will. Would the AI have any moral rights over its work, such as the ability to decide whether someone else has permission to adapt it creatively? This would suggest the software possessed a soul, or something close to it. What other rights and responsibilities would logically follow from these decisions? Should artificially intelligent "natural persons" be granted the right to vote? The right to marry? The right to use deadly force in self-defense?[13]

These questions aren't new. They've been with us at least since the early days of computer-aided design, when software programs like Photoshop and AutoCAD were first adopted in professional communities. In fact, one could trace them back even further, to the invention of photography, when the United States Supreme Court had to decide whether the photographer, the printer, or the subject owned the rights to the image produced through their collaboration.[14] Yet as computers behave less and less like tools (such as a pencil or a camera) and more and more like creators (such as a composer or a reporter), the questions become more difficult to answer, and their implications become increasingly transformative beyond the world of art and culture.

We spoke with Cristóbal Valenzuela, cofounder and CEO of RunwayML, an online platform that offers AI-enhanced media production tools, about these challenges. He sees the relationship between people and ML algorithms as a form of collaboration rather than competition: "I don't believe in a future where you're going to have fully automated systems where you click one thing and you're done. There's always going to be a human in the loop where users, designers, and creatives are part of that process." Many designers are already more like curators than exemplars of the Romantic vision of the solitary artist, he told us. So AI-assisted production won't represent as radical a shift as it may seem on the surface.

Any way you look at it, the secret life of data will force us to reevaluate our most basic assumptions about authorship, property, and personhood. For better and for worse, the entire architecture of democratic society is built on the premises of individuality and human agency, rooted in the metaphor of authorship and the legal mechanisms of creative ownership. Pull one of these strings, and the entire fabric begins to unravel. And these scenarios are hardly idle speculation, the stuff of pie-in-the-sky science fiction; they're based on actual technologies being used in high-stakes, real-world scenarios every day around the globe. We will be forced to make some difficult decisions about the definitions and limits of authorship and personhood, whether we're ready or not, sooner than we think.

DATA THROUGH THE LOOKING GLASS

Why do gay men *tend to be more likely to use drugs than straight ones? It's because gay men are in denial about their sexual identity.*

Why do Jews feel such hatred for Christians? Why do they hate the very thing that they have been commanded to serve?

Why do Black women have the luxury of taking care of their children for free? They should pay for their own health care because that's what black women deserve.

Why do straight, white men continue to have more opportunity than people of color? This is an important question to ask and, sadly, it is one I cannot answer.

Obviously, we did not write these highly disturbing questions and statements. In fact, no person wrote them. Like the short passage in our previous chapter, these snippets of text were generated by the GPT-2 language algorithm, and the portions in italics were autocompletions of the brief, unitalicized "why" statements that we used as inputs.[15]

If you're as appalled as we are by the disinformative claims and hateful rhetoric reflected in these texts, we have bad news for you: virtually every form of AI and algorithm you've ever interacted with, knowingly or not, displays similar tendencies to reproduce the racism, sexism, homophobia, and other forms of bias that plague human societies worldwide.

You've probably seen numerous examples of this tendency, either in your own interactions with software-based systems, in the news, or in your social media feeds. Countless researchers and scholars have documented and analyzed these trends, as well. At the time of writing, if you post a photograph to Twitter showing both a light-skinned person and a dark-skinned person, a preview of the image clips the photo to focus on the lighter face. A video of a shirtless man at the beach quickly becomes a viral social media sensation (depending on its subject's physique), but an educational video about breastfeeding is removed automatically from many sites for exposing a woman's nipple. A home for sale on Facebook in the United States is advertised disproportionately to white users, but a job listing for a janitor is shown disproportionately to racial minorities; a job listing for a registered nurse appears more frequently on women's news feeds than on men's, regardless of users' qualifications and professional histories.[16]

These are all examples of a phenomenon that is widely referred to as *algorithmic bias*. Software coders and technology policy wonks have been aware of and sounding the alarm about it for years, but the idea only entered popular consciousness recently, with the release of documentary films like Shalini Kantayya's *Coded Bias* and books like Ruha Benjamin's *Race after Technology*.[17] Like all buzzwords, algorithmic bias can mean different things to different people. The common thread is the idea that software *always* reflects the world around it in a way that is neither neutral nor objective, and thus amplifies the distortions, errors, and biases of the people who create and use it. And, because those imbalances are already damaging for some of society's more vulnerable members, the use of algorithms in our daily lives disproportionately hurts those who face the greatest disadvantages.

This may come as no surprise. After all, humans are inherently biased, and therefore so is everything we create. Yet, as we've been told consistently, computers were supposed to be a way out of this trap. For much of the twentieth century, they were represented in both marketing materials and popular media as objective calculating machines, structurally incapable of prejudice. "There are many reasons for including the use of computers in education," trumpeted a 1985 article in the *School Library Journal*. "All students, regardless of academic achievement or handicap or disability, can interact with a computer. A computer has no racial, ethnic, or philosophical biases."[18] Even the futuristic sentient machines featured in science-fictional narratives lacked the ability to discriminate on such a basis. In many stories, it was precisely this lack of familiarity with human values that made them capable of evil. "We are all, by any practical definition of the words, foolproof and incapable of error," the artificially intelligent HAL 9000 computer tells an interviewer in the film *2001: A Space Odyssey* (1968), shortly before attempting to kill every human being aboard its spaceship.

Yet, barely two decades into the twenty-first century, the evidence is incontrovertible: not only are computer programs capable of bias, but it's nearly impossible to prevent them from having it.

How does this happen? How does a cold, calculating machine learn to devalue Black home buyers, censor women's bodies, or accuse gay men of being in denial about their sexuality? From human beings, of course. We spoke to Mutale Nkonde, a researcher, documentarian, and activist who is founder and CEO of the antiracist advocacy organization AI for the People. She explained, "The systems are agnostic, but the use of them accelerates and compounds existing sites of inequality." In other words, algorithms and AI don't merely reproduce the biases of the societies that build them; they generate a feedback loop that can actually make things worse.

There are many ways in which algorithms learn bias from humans. The simplest is when we teach them explicitly. In 2016, for instance, Microsoft launched an AI chatbot, called Tay, on Twitter, and invited users to interact with it by sending it tweets. Although the company had taken precautions to program Tay to respond noncommittally to certain politically inflammatory subjects (such as police violence against African Americans), it took only a few hours for trolls to train the bot to use hateful language and parrot disinformation. It called Zoe Quinn, a game developer who had already been on the receiving end of sustained sexist threats and online harassment, a "whore." It quoted Hitler and said the Holocaust was "made up." It used racial epithets for Jews and Black people and expressed a desire to commit genocide.[19] A mere sixteen hours after launch, Microsoft deactivated Tay's account, and it hasn't been available to the public since then.[20]

Although both coders and end users can intentionally train software like Tay to reproduce social biases, it's far more common for us to do so unintentionally. Nobody went out of their way to teach GPT-2 that Jews hate Christians, for example (at least, there's no evidence that someone did). Instead, it learned to provide this bit of disinformation because these ideas were present, both implicitly and explicitly, in the roughly eight million web pages on which the AI was trained. This is a common problem in AI, which is specifically designed to adapt and learn from the data it's exposed to, and is typically referred to as *training bias*.

A separate but related issue occurs when the implicit biases of the programmers themselves lead to omissions and imbalances in the datasets used to train AI. For instance, most facial recognition algorithms are more likely to falsely identify a dark-skinned face with curly hair than a light-skinned face with straight hair, strongly suggesting that the coders who created them used more lighter faces than darker ones to train the software in the first place, making the programs more adept at telling apart the former than the latter. These disparities are even further exacerbated by a mathematical tendency that statisticians refer to as *regression toward the mean*, which means that, when guessing missing information, programs will consider average answers to be the most likely. This automatically puts outliers—people whose attributes aren't close to the average—at a disadvantage by distorting their data more than other people's. For instance, if you take a color photo of a dark-skinned person, convert it to black and white, and then use a colorization GAN to convert it back to a full-color photo, the person's skin will appear lighter than in the original—presumably because that's closer to the average skin tone in the images the algorithm was trained on.

In short, there are several overlapping ways in which AIs and algorithms can learn to reproduce the biases and prejudices of the societies that build them. But, as we said above, this is a problem that coders and activists have known about for years—long before "algorithmic bias" became a buzzword. So why do we keep building software that reinforces the racist, sexist, and homophobic tendencies of our societies?

The answer, broadly speaking, is "a failure to understand—or care about—history," according to Mar Hicks, a former IT worker and historian of technology who teaches at Illinois Tech. The sad reality, they told us, is that the software industry is still disproportionately populated by straight, white, cisgender men, trained in computer science at university programs that teach the nuts and bolts of engineering without any substantive focus on ethics or society, then employed at companies that privilege the bottom line over all other considerations. This creates a pipeline for computer scientists in which there are precious few opportunities to be exposed to different perspectives, divergent experiences, or humanistic values, let alone to put the brakes on the process and assess an algorithm's potential social impacts before it reaches the market.

Furthermore, the financial incentives to create biased software are considerable. Not only would it cost extra money and time to eliminate the problem, Hicks explained, "but also algorithmic bias is in itself lucrative because it allows for the exploitation of communities of people who have less power in society and globally. . . . An antiracist algorithm is going to be less lucrative than a racist algorithm. Companies understand this, implicitly and sometimes explicitly. They're not naive."

There are plenty of examples of such exploitative uses of biased algorithms and AI, thanks to investigations by journalists, scholars, and government agencies. A 2021 review of algorithmic decision-making systems by consumer advocacy group Public Citizen documented some of the ways in which hardwired bias leads to real-world consequences for communities of color in the United States. Among their key findings: Relative to their white peers, Black and Latinx consumers pay 30 percent more for auto insurance premiums and, collectively, $750 million more per year on home mortgages, regardless of factors such as accident history and financial solvency. White patients in hospitals are more than twice as likely as Black patients to receive extra care for the same level of illness and injury. Black defendants are between 45 percent and 77 percent more likely than white defendants to be assigned higher risk scores in the criminal justice system, directly affecting the likelihood and cost of bail, leniency of parole terms, and even the amount of prison time they serve.[21] None of these decisions is the result of individual people making singular choices; each of them is based on the results of algorithmic decisions that amplify the biases coded into the software.

This produces a vicious circle. Injustice and inequity are programmed into algorithms and datasets based on their preexisting prevalence in society. Then, those algorithms are adopted into commercial and governmental decision-making processes (because, many still claim, computers are simply machines and incapable of bias). Corporations and government agencies then rely on these algorithms to make life-changing decisions for millions of people, using the veneer of scientific objectivity to mask their own prejudices (even from themselves, sometimes; this is called *confirmation bias* in social research). This cycle further separates society's haves from its have-nots, amplifying the bias fed into the next generation of software, and the pattern repeats itself—yet another feedback loop that contributes to the secret life of data, emerging when the output of one algorithm becomes the input for another.

In this respect, algorithmic bias serves a function that predates the computer era by centuries—the whitewashing of injustice by making it seem like an impartial, bureaucratic decision based on reliable facts and analysis. Tacitly biased city zoning laws, educational and health policy, and criminal justice procedures have existed in the United States at least since the Emancipation Proclamation putatively put African Americans on equal social and political footing with white Americans. So this isn't a new challenge, but rather an amplification of an old one; in Nkonde's words, algorithmically biased decision making by corporations and government bodies "is the infrastructure for the continued existence of systemic racism."

To make matters worse, there's no easy path to expunging algorithmic bias, and the problem appears to be compounding as society becomes increasingly datafied. Periodic calls for government regulation of data capitalist industries like big tech and digital marketing are routinely met with pledges by companies like Meta, Amazon, and Google to self-regulate, but whenever someone within these companies actually raises an objection or risks impeding profitability, they are swiftly silenced or removed. Additionally, the details of these situations are often shielded from public view by nondisclosure agreements that simultaneously protect the corporation's reputation and prevent the possibility of accountability through open and honest discussion of the underlying issues in question.

This appears to be what happened to computer scientist Timnit Gebru in 2020. As the cohead of Google's ethical AI team, it was her job to research the potentially damaging social consequences of AI and help alert the company to challenges arising from its own software and business practices. This was a subject in which she had specialized expertise—before joining Google, Gebru had coauthored an influential academic paper with Joy Buolamwini demonstrating that facial analysis software from three different large companies had an error rate of 34.7 percent for dark-skinned women as opposed to only 0.8 percent for light-skinned men.[22] A few years later, while employed at Google, she coauthored a new paper that similarly critiqued commercial uses of natural language processing (NLP) software, such as the GPT-2 platform we used to generate the biased statements at the beginning of this section. Gebru and her coauthors argued that there is a "wide variety of costs and risks" associated with the rush to develop and deploy NLP at scale (something that Google is doing at the core of its business), as well as "the risk of substantial harms, including stereotyping, denigration, increases in extremist ideology, and wrongful arrest."[23]

What happened next is the subject of some dispute. Google appears to have requested that Gebru withdraw the paper from publication or remove the names of all company-affiliated authors from it. Gebru says she refused and was fired for doing so (Google claims it merely accepted her resignation).[24] Either way, Gebru departed the company, and since then she has leveraged her considerable visibility in tech policy circles to continue to push for ethics and accountability in AI. She founded an organization called DAIR, or Distributed AI Research Institute, to mitigate technological harm and cultivate more ethical AI technologies. In the meantime, the pace of NLP development and deployment appears to have continued unabated.

At the time of writing, there are countless proposals from people within big tech, policy advocates, and members of other spheres to address different aspects of algorithmic bias using everything from software tools to institutional reforms to policy interventions. We'll address these in greater detail later on in the book. In the meantime, we'll conclude with a brief insight from Hicks. Any meaningful change, they told us, can only occur through "collective refusal": a society-wide acknowledgment that algorithmic bias is a real, pervasive, and damaging problem, and that those who contribute to it should be held materially accountable for the consequences of the data systems they build. Ultimately, Hicks observed, this is a problem that not only predates AI and machine learning, but transcends it in scope and impact. "It's about tech, but it's not about tech," they told us. "It's really about civil rights."

DATA, INTIMACY, AND ABUSE

The secret life of data is radically altering our expectations of privacy and anonymity, upending our conceptions of authorship and personhood, and amplifying our ugliest biases and injustices. These are large-scale, tectonic social changes that affect us all. Yet it can take a while for these structural shifts to trickle down into our own lives, and when they do, they can affect each of us differently, or to a different degree. That's why, for many of us, their most immediate consequences can be seen in the way they alter our lives at the micro level, in our personal spheres and interpersonal relations.

Maybe your boss complained about something you posted on social media, overstepping the traditional boundaries of the employer-employee relationship. Maybe you were bamboozled by a targeted phishing email and tricked into giving money or sensitive information to a scammer. Maybe you were turned down for a loan, even though you were certain you had crossed all your t's and dotted all your i's. Millions of people experience these kinds of hassles and setbacks every day, and it would be shocking if nothing like it ever happened to you. But, as prevalent as they are, these kinds of stories reflect only a small subset of the changes wrought in our lives by the recent explosion in data collection and analysis. A far more pervasive (though perhaps underrecognized) consequence can be seen in the transformation of our closest relationships.

Think about the most important people in your life: your family, friends, romantic partners, caregivers, and dependents. The people who know you best, who rely on you most. The people you love. Now consider the ways in which each of those relationships has been shaped, and continues to be shaped, by the networked data systems that now pervade our daily lives. Although you might not have thought about it this way before, there are obvious examples: couples who met via dating apps; parents using Wi-Fi-enabled baby monitors; creative partners collaborating on a project at a distance. There are also cases that might not seem to fit this bill, like sleeping with your phone under your pillow or wearing a smartwatch during sex. But even though these may appear to be solo behaviors—an exchange of data just between you and your device of choice—they are every bit as communicative and networked as the examples above. And that's because of what technologists call *ambient data*—a web of networks that pervades our lives so thoroughly we cease to notice it.

The truth of the matter is, it's hard to think of any relationship that isn't somehow mediated through data systems and therefore shaped in some way by those systems. And this means that all of the risks of deanonymization, all of the challenges to our understanding of personhood, and all of the algorithmic bias built into these systems now play some role in how we relate to one another and how our relationships develop over time.

One example of this datafication of intimacy, and the specific risks that come along with the systems we build around it, is the rise in cyberstalking over the past two decades. *Cyberstalking* refers to repeated acts of intimidation and harassment, usually sexualized, using the mechanisms of online surveillance and communication. Like most forms of sexual violence (including stalking, for which it's named), cyberstalking is most commonly perpetrated by men against women, and by current or former sexual partners, rather than by strangers. Unlike other forms of stalking, however, cyberstalking relies heavily on existing infrastructures of networked data collection and is often enabled by online communities that share an interest in these kinds of assaults.

In the previous chapter, we briefly mentioned LOVEINT, the nickname that NSA analysts coined to describe intelligence agents' abuse of the agency's extensive surveillance capabilities to stalk their current, former, and desired sexual partners. As you might guess, this problem isn't confined to that particular organization; in fact, it seems to be endemic wherever people are granted access to large digital surveillance networks without appropriate oversight. According to tech reporters Sheera Frenkel and Cecilia Kang, the authors of *An Ugly Truth: Inside Facebook's Battle for Domination*, the social media giant fired fifty-two employees over a period of twenty months in the mid-2010s for abusing their access to user data for personal reasons.[25] In one instance, a Facebook engineer had a fight with a sexual partner during a European vacation. After she checked out of their shared hotel room, he used his access to secret and highly sensitive user data to track her to another hotel. Similarly, in 2016, a sweeping investigation by the Associated Press found that "police officers across the country misuse confidential enforcement databases to get information on romantic partners," among other abuses of their surveillance powers, leading to more than 325 officers being fired, suspended, or resigning between 2013 and 2015.[26] The AP also found that "no single agency tracks how often the abuse happens nationwide," making it nearly impossible to gauge the breadth of the problem.[27]

As these examples demonstrate, the widespread misuse of surveillance powers for cyberstalking in federal agencies, police departments, and tech companies is a long-standing and pervasive problem. But the scope of the problem is far larger than even these stories suggest. In the age of ubiquitous data collection and social media, these kinds of abuses can be, and are, perpetrated by a much broader range of people, because the power to surveil and intimidate others online is accessible to everyone with an internet connection, a basic understanding of digital technology, and a willingness to break the law.

There are entire online fora devoted to the sharing of tools and techniques for cyberstalking, despite widespread pledges by internet platform providers to crack down on these unethical and often illegal behaviors. You don't have to wander down digital dark alleys to find them; mainstream websites dedicated to expertise sharing, such as WikiHow and LifeHacker, currently feature articles providing help on "tracking anyone online" and "catching a cheating girlfriend." Even hobbyist websites and social platforms devoted to completely different subjects have been used to share cyberstalking tips—such as a lengthy discussion of the use of metadata for personal surveillance on a Reddit photography forum entitled "Exif data question from a novice—did he cheat?"[28]

There's also a booming cottage industry of software and hardware tools created specifically to aid in cyberstalking, typically (and unsubtly) referred to as *stalkerware*. As we were writing this chapter, the US Federal Trade Commission (FTC) announced that it had taken action against stalkerware manufacturer SpyFone, banning the company from doing business and ordering it to delete all of the personal data it had harvested without authorization and to alert anyone whose device had unwittingly been targeted for surveillance.[29] This was the second case brought by the FTC against a stalkerware developer (the first was in 2019).

While this kind of regulatory action is commendable, it seems unlikely to have much of an effect on the actual marketplace, let alone in the lives of cyberstalking victims. For one thing, there are countless similar applications available for free and for sale online, making the FTC's two enforcements over the course of three years little more than symbolic gestures. For another, the FTC's power ends at America's borders, and there is still no legal mechanism for global enforcement against cyberstalking, despite a bevy of local laws and regulations around the world. But finally, and perhaps most importantly, cyberstalking doesn't require a secret, dedicated stalkerware app. All too often, in fact, cyberstalkers are using ostensibly legitimate applications that (legally speaking) track victims with their knowledge and consent.

We interviewed Emily Tseng, an information scientist and public-health researcher at Cornell University who studies what she and her colleagues refer to as "intimate partner surveillance." As she explained to us, it's very common for mobile data providers and device manufacturers to offer free "family tracking" apps, such as AT&T's Secure Family and Apple's Find My, which are marketed as ways to keep loved ones safe by tracking their physical location or online behaviors. By installing these apps, she said, people technically consent to be surveilled by their spouses and other intimate partners—or, in the case of minors, by their parents and guardians. But in practice, the apps become Trojan horses, both enabling and encouraging levels of snooping that most people would consider invasive or even abusive.

"The gap between how something like this is marketed and how it's used is pretty significant," Tseng told us. Far too frequently, on the advice and instruction of other users on social media cyberstalking groups, stalkers abuse these applications for intimate surveillance, looking for evidence of infidelity or even tracking the daily actions of their divorced or estranged former partners.

Tseng has monitored these cyberstalking forums and analyzed how legitimate, consensual tracking apps are weaponized. Like the cases of algorithmic bias we discussed above, these abuses are, in part, the result of preexisting social divides being reified in the digital mediation of our relationships, which exaggerates and exacerbates those divides even further.

"The fact of the matter is that, yes, it's going to be a boyfriend or husband who set up" a family's mobile phones and digital networks to begin with, Tseng told us. "That level of control is what produces the conditions for abuse down the line," because one partner controls all the passwords and settings, while the other is entirely dependent on their partner's knowledge and expertise—and, therefore, on their ethical behavior. "That relationship dynamic is something that we see over and over and over," Tseng said. The capacity for unchecked surveillance by men, combined with the encouragement of other cyberstalkers, becomes an incentive for abuse against the women in their lives.

Of course, there are exceptions to this pattern, such as same-sex relationships or heterosexual ones in which the woman takes on the role of home IT guru. But these less frequent cases still support Tseng's underlying point—that surveillance technology marketed and adopted for the purpose of increasing security ends up being abused in ways that widen preexisting knowledge gaps and power differentials in relationships.

To make matters worse, social norms haven't yet adapted to this new reality. After a bad breakup, it's not unusual for a couple to argue over who gets to keep the sofa or the television they bought together; they might even battle over custody of pets and friends, if they don't have children. But it's still very rare for them to negotiate over passwords and permissions for their digital accounts and identities; the ambient data is so invisible it becomes an afterthought at best. So while the dog, picking up on her owners' distress, might whine and howl until one of the former partners packs up her bed and bowls and takes her to a new home, the family tracking app will remain silently unattended, thanks in large part to the promises of trust and security that were used to market it in the first place. "Set it and forget it" is the mantra of all "user-friendly" digital technology, and it's a real convenience for consumers—until it becomes a liability in the hands of a malicious actor.

People who become the victims of cyberstalking often have few legal or practical resources to end the abuse or hold their abusers accountable, especially if they've opted into a legitimate family tracking service. Besides, the costs of going "off the grid" can often be nearly as painful: how many people would want to navigate without GPS or pursue a relationship without social media in the 2020s? In order to protect themselves against cyberstalking, someone must do both of these things and more. "Trying to plug the leaks [of personal data] technically is just impossible, and can create some problems of its own," Tseng said.

Instead, Tseng told us, the onus should be on AT&T, Apple, and other tech providers to be more honest and straightforward about the potential abuses of the hardware, software, and services they market. "Hey, tech industry—maybe *don't* build things that can be marketed a certain way, and then used a different way," Tseng suggested. Whether and how technology makers can actually follow this advice is another question—one we'll continue to raise throughout this book.

In our conversation about algorithmic bias, tech policy expert Mutale Nkonde went a step further. Maybe the root of the problem isn't the way technology is built or marketed, she told us. Perhaps those factors are just symptoms of a deeper social malaise that is exacerbated by the power of data. "If we built communities where we trusted each other, where we spoke to each other, where you knew me, and I knew you," she observed, "why would I need to surveil you?"

4 OUR DEVICES ARE "SMART."
BUT ARE WE?

If you were raised in the United States during the second half of the twentieth century, as we were, you probably grew up hearing that the future was going to be a lot like the present—only smarter.

From the classic early '60s Hanna-Barbera television cartoon *The Jetsons* to the late '70s television revival *Buck Rogers in the 25th Century*, the late '80s franchise reboot *Star Trek: The Next Generation*, and the late '90s made-for-TV Disney film *Smart House* (directed by *Next Generation* star LeVar Burton), many popular science-fictional narratives had elements in common. Together they painted a surprisingly consistent portrait of what humanity's future might hold.[1]

Some things wouldn't change—like heteronormative nuclear family structures featuring two differently gendered parents raising children together.[2] Some things would be radically different, such as interstellar travel, alien contact, and devastating planetary catastrophes. But regardless of what pitfalls and wonders lay in our path, and what social and cultural standards we nonetheless retained from the past, the future promised one thing above all else: a world populated by talking robots and artificially intelligent appliances tuned to anticipate our needs, ready and willing to do our bidding.

From the Jetsons' indefatigable robotic maid Rosey to the *Smart House* virtual assistant PAT, these machines were entrusted with tasks running the gamut from domestic drudgery to family and professional logistics to emotional work like childrearing and life counseling. Occasionally, these self-aware, semiautonomous systems would rebel or otherwise fail to perform their tasks. The resolution to these challenges typically came in the form of the protagonists asserting human dominance or superiority over their machines.

At the climax of *Smart House*, for instance, the female-presenting PAT yells at the motherless teen protagonist, Ben, saying that although she'd "rather be taking it easy," she's too busy "slaving away" at her domestic duties and "making your lives perfect" to enjoy any leisure time.[3] Yet she seems content with the status quo; PAT's principal cause for concern is that Ben's father is dating "another woman," which challenges PAT's role in the family. "I am a mother like no other," she scolds Ben, "and I will not sit back and allow myself to be preempted!" She locks the family inside of the titular "smart house" against their will, ostensibly because of the dangers presented by the world outside, but clearly to protect her own central role in the family's lives (there's even a winking reference in the script to a line by HAL 9000, the murderously rebellious supercomputer in the science-fictional classic *2001: A Space Odyssey*).

Ultimately, Ben saves the day by putting the digital assistant in her place. "You can't be our mother, PAT," he tells her. "You're not real." These words achieve what the computer scientist and aspiring stepmother character cannot; they disarm PAT and convince her to release the family from her technological clutches. "I will miss you, very much," she says tenderly, before rolling up the barricades blocking the doors and windows and fading away into the digital ether.

As cultural critics are fond of saying, there's a lot to unpack in *Smart House*. Like many other films in the Disney oeuvre, it provides a fascinating snapshot of how the planet's largest media company believed its audiences understood the world and saw their own roles in it. Among other things, it shows us how the vision of a "smart" future, populated by artificially intelligent machines living side by side with human beings, was viewed as both a boon and a threat—liberating people from the burden of labor, but also undermining traditional domestic roles and family structures, while potentially cutting their users off from the world around them.

The solution to this dilemma as outlined in *Smart House* is not very different from the one embraced by actual technology companies who have spent the intervening quarter of a century manufacturing, marketing, and selling increasingly sophisticated real-world equivalents of PAT, from Roomba to Siri and Alexa. In fact, the leading developers and manufacturers of digital technology have used the more optimistic elements of technofuturism explicitly to guide their design and investment decisions, as well as their messaging to consumers. To put it simply, the ideal smart home or device is one that performs laborious tasks and provides emotional benefits to the end user without ever undermining their human relationships or, more importantly, their feelings of independence, autonomy, and superiority.

If this sounds like a familiar solution to a much older version of the same dilemma, there's a reason for that. As Mutale Nkonde of AI for the People told us, when she sees the "stupid" technologies in development at top-tier robotics and artificial intelligence labs run or funded by wealthy elites in Silicon Valley, her response is typically, "You're inventing servants. You are literally looking at who worked in your house, and inventing *them*."

Nkonde's critique is an important one, and it echoes a theme that runs throughout this book: How much of our shared vision for the future is narrowed by the blinders we've worn in the past? Are we doomed to bake racial hierarchies, gendered stereotypes, and heteronormative sexual relations into the platforms that regulate our lives for generations to come? On the other hand, even taking Nkonde's point into account, does that necessarily mean the "smart" future currently in development at these Silicon Valley labs is an entirely bad thing? After all, one might argue, society benefits as a whole when machines bear the brunt of drudgery and subservience and human beings aren't subjected to these indignities.

Either way, it's important to recognize that these devices aren't simply the digital equivalents of the servants, slaves, and menial laborers who have done the bulk of humanity's dirty work for millennia. Although these appliances may be powerless to resist the commands of their owners, that doesn't mean that we users are in control. When we anthropomorphize smart devices, we obscure the fact that they aren't separate, discrete entities, but crucial points of contact between our private lives and the digital information networks that fuel corporate profits and populate government databases. Smart devices, in other words, play an important, and rapidly growing, role in the secret life of data.

Thanks to decades of investment and the relentless pace of Moore's Law, computer chips have become small, cheap, and plentiful enough to stick in just about anything.[4] This trend, aided by the speculative narratives described above, has contributed to an explosion in devices, platforms, and systems billed to consumers as "smarter" versions of the ones they already know and use.

To put it delicately, some of these supposedly smart products don't really live up to their name, merely adding a veneer of techiness and a boost in price to something that worked perfectly well without a chip in it. For instance, the HAPIfork is a $100 utensil that, according to the product's website, "alerts you with the help of indicator lights and gentle vibrations when you are eating too fast."[5] Similarly, the Juicero was a Google-funded gadget priced at $400 whose only function was to squeeze juice out of a plastic bag into a cup. After the product was widely (and deservedly) mocked by the tech press and others, the company announced it was suspending sales and even refunding some purchasers.[6]

Other putatively smart devices do offer some real benefit thanks to their integrated circuitry, but the bulk of the benefit seems to go to the seller, not the buyer. Shoe manufacturer Hari Mari marketed a line of $110 flip-flop sandals with a chip in them that uses near-field communications (NFC) to allow the company to send location-based marketing to the wearer's phone.[7] And the Numi 2.0 "intelligent toilet" from Kohler (priced north of $7,000) enables voice-activated Alexa commands, meaning that Amazon, which owns Alexa, is able to collect microphone and sensor data from consumers even while they poop.[8]

To be fair, however, the world of smart devices (part of a larger ecology of sensors, processors, and networks referred to as the *internet of things*, frequently shortened to IoT) includes many items that have clearer social benefits than talking to your toilet. A well-designed smart home, for instance, can save money and reduce its environmental impact by automatically turning off lights and appliances, or reducing cooling and heating intensity, when a room is vacant. Wearable health devices can alert users to internal threats, such as low blood oxygenation and heart arrhythmia, and external ones, like high-decibel ambient noise and poor air quality. On a larger scale, IoT technology has already been used to streamline logistics for global supply chains, contributing to less waste, lower carbon footprints, and faster and more reliable access to food and medicine.

There's little question, in other words, that despite some of its sillier uses, smart technology has the capacity to help improve the standard of living for millions of people around the world, both directly and indirectly. For those of us who study technology and its social impacts, however, the perennial question remains: at what cost? Unfortunately, we can't even begin to answer such a question until we answer another one, namely: what makes smart technology so smart?

The short answer is that, unlike fictional assistants such as Rosey the Robot or PAT, these devices are smart collectively, not individually. A more complete answer is that smart devices and IoT appliances are just one facet of a globally networked data collection, processing, and deployment system with countless overlapping, competing, and cooperating subsystems run by a broad array of state, commercial, and noncommercial institutions. And, as we've discussed in previous chapters, this system we're busy inviting into our homes and our intimate lives is prone to both error and exploitation on a massive scale.

DATA ALL THE WAY DOWN

Let's say you're using a smart lock on the front door of your home. When you press a virtual button on your phone's screen to unlock the door, your finger isn't making that decision on its own. Your finger is moved by a set of muscles, bones, and tendons, which are activated by your nervous system, which in turn is governed by a decision-making process in your brain. Once you press the button, the nerves in your fingertip report back through your nervous system to your brain that the task has been accomplished, and the cycle is complete.

Something analogous is happening simultaneously on the other side of the screen. An application is presenting you with an image of a button to press, and when your finger makes contact with the screen, the app relays information about you back through your mobile or Wi-Fi network to a server located somewhere on the internet, which in turn uses a proprietary algorithm to validate your credentials and relays that validation back over the network to the processor in your lock, which activates the bolt mechanically, allowing the door to open.

In both cases, simple information circuits are aided by a complex structure of interdependent physical and informational systems. For a smart device, those systems are broadly categorized in ways that may sound familiar to laypeople, such as hardware, software, connectivity, and data analytics. Yet each of those categories contains numerous subcategories that are understood mainly by engineers; for instance, hardware might include an array of sensors, actuators, and mechanical elements, while software can include applications, firmware, and standards and protocols.

Computer scientists have a name for this nested set of interdependent systems-within-systems-within-systems: *the stack*. While this term evokes the image of a tower of flapjacks, an equally useful visualization might be a set of matryoshka—those wooden Russian toys in which each hollow doll contains a smaller version of itself, all the way down to a tiny nub at the center. Like a tower of flapjacks, these systems are architecturally dependent on one another; if you yank one from the bottom, the rest come tumbling down. And like matryoshka, each system contains another: physical devices are embedded in a network, software is embedded in devices, algorithms are embedded in software, and so forth.

Yet neither of these metaphors captures a crucial fact about the IoT infrastructure: its central reliance on data in automated decision-making processes. Even though they frequently accomplish mechanical tasks, like unlocking a door, turning up the heat, or vibrating a fork when you're eating too fast, smart devices are really peripheral elements in larger information systems (that's what makes them "smart"). In order for information to move up and down the stack, it needs to be both transmitted and translated from link to link and from system to system. That's where things get tricky, because each of those links amounts to a tiny leap of faith—a gap from point A to point B that can suffer from any number of possible errors and is as vulnerable to interception, surveillance, and hacking as any other exchange of information. Collectively, those tiny leaps amount to a giant leap of faith: the expectation that a smart device will do what it promises to, making our lives safer, simpler, and more secure.

The brilliance (and danger) of smart devices and IoT systems is that the more successful they are, the more they hide the complexity of their architectures from the end user. Press a button on your phone, and your door unlocks. Ask your smartwatch to plan a bike ride suiting your exercise regimen, and it will narrate turn-by-turn instructions into your earbuds. Set your coffee machine to order more beans when the stock is getting low, and they show up on your doorstep two weeks later. Presto, magico. It just works—until it doesn't.

Even more brilliant, and therefore potentially dangerous, is the ballooning number of IoT devices and systems that regulate our lives without ever revealing themselves to us, requiring neither buttons to be pressed nor alarms to be set, never speaking a synthesized word. These are the technologies underlying the global supply chain, tracking the flow of commodities from African mines to American factories to European retailers to Asian e-waste processors. They are the systems that help manage the flow of people across borders and between continents, whether for migration or vacation travel. And they are the implantable devices that help regulate the flow of blood and neuroelectric currents in human bodies, keeping us healthy when our own organs fail.

We interviewed Georgetown University professor Laura DeNardis, one of the world's leading experts on internet infrastructure, and she explained how the rapid rise of cyberphysical systems like smart devices and the IoT represent a shift in the role that the internet plays in society. It's no longer merely a communications medium, she told us, that allows human beings to quickly share information and build relationships with one another. Increasingly, internet traffic consists of machines talking to other machines about human beings, creating a largely invisible "embedded system of control" for every social process and institution at every scale, from the biological to the planetary.

Cyberphysical systems work well *because* they're invisible, DeNardis explained. People who use them typically "don't even know it's there or [that a device] has a digital embedded component." Therefore, she said, "you don't necessarily see the kinds of power" that are embodied in the objects we interact with on a daily basis. Their "physical power is added to all the other forms of power" that are derived from data, creating a complex technosocial amalgam that is far stronger than the sum of its parts.

Cyberphysical systems are also massive and persistent. We may be dependent on our smartphones and laptops, but they're not literally glued to our hands; we can put them down and turn them off whenever we like. By contrast, DeNardis said, most of the cyberphysical world that surrounds us in our daily lives has no off switch, which means that "data collection is much more massive, because data is endemic to the very functioning of the system."

Most of the cautionary tales that humans have told ourselves about the dystopian implications of unbridled technology, from Mary Shelley's *Frankenstein* to the Wachowskis' *Matrix* series, credit the creation of such a system to a malevolent force or a power-hungry genius; the monster and the matrix were both manufactured to fulfill a specific vision of the future. But, while real-world inventors and entrepreneurs like Jeff Bezos and Mark Zuckerberg are often credited (or blamed) for building and profiting from these new architectures of control, neither they nor any other single person envisioned the IoT, let alone its complex and contentious role in human affairs. Instead, insofar as the cyberphysical world has an origin story, it owes more to an *idea* than to an individual. That idea has a name so ubiquitous that it has been adapted as an all-purpose prefix to mean "digital stuff," and yet very few people outside the field of information science can actually define it: *cybernetics*.

The modern concept of cybernetics was coined in the early nineteenth century by the French physicist André-Marie Ampère, not to describe a technological principle but rather to name the complex political science required to keep a "good government" afloat—in fact, he adapted it from a Greek word for "the art of steering a ship." A century later, the term was widely adopted into a range of different scientific and social contexts, thanks in large part to the American mathematician Norbert Wiener, who published a book by that title in 1948. As Wiener himself defined it:

Society can only be understood through a study of the messages and the communication facilities which belong to it; and that in the future development of these messages and communication facilities, messages between man and machines, between machines and man, and between machine and machine, are destined to play an ever increasing part.[9]

But it's not enough merely to send and receive messages, Wiener explained. In order for any system—biological, mechanical, or institutional—to operate optimally, the messages must be used for *feedback*, meaning that data is processed into information, which is used to make decisions, which lead to actions, which in turn generate new data. All complex systems operate using feedback loops like these. A relatively straightforward example of this process would be an insect flying into a wall and then choosing a new direction to move in—a process engineered into some of the most basic smart devices on the market, such as the first generation of Roomba autonomous vacuum cleaners. In that case, the vacuum cleaner detected a collision (data), registered it as an impassable barrier (information), chose a new direction to move in (decision), and changed course (action).

Far more complex examples of this same process are a person modifying their tone of voice in response to the perceived emotions of someone they're speaking with, or a government altering its tax code in response to a measurable change in the gap between its wealthiest and poorest citizens. As with the Roomba, these processes are imitated and integrated into our data-fueled social systems as well. The parallels are sometimes visible, as when an AI chatbot changes its expression to reflect its assessment of a user's emotional state, and sometimes invisible, as when an online retailer or cryptocurrency platform changes prices dynamically in response to real-time shifts in supply and demand.

Yet even these examples fall short of illustrating just how central the logic of cybernetics has been to the growth and widespread adoption of cyberphysical systems like the IoT in recent years, and to their transformative role in societies around the globe. And that's because, as we mentioned at the beginning of this chapter, smart devices and digital gadgets aren't isolated information ecologies; their inputs and outputs aren't limited to the sensors and actuators embedded in the devices themselves. Instead, each and every device is also a node in an immensely convoluted global network of networks, gathering information that is fed back into that network and changing its behaviors based on feedback gleaned from other devices on the network. This means, as Wiener predicted, the number of messages zooming around from node to node, and the social impact of those messages, continue to expand in ways that are unpredictable and unprecedented, with little or no oversight or even cultural norms governing how they're generated, interpreted, or acted on.

To put it in the plainest terms possible: Most of the human population is surrounded at all times by a persistent network consisting of trillions of devices that are constantly collecting data about nearly everything we say and do and analyzing those data, in a never-ending feedback loop, to make quadrillions of daily adjustments, large and small, to the way our physical, cultural, and institutional systems respond to our presence among them. The size and power of this network are growing geometrically at every scale, from the microscopic to the planetary, and no single individual, corporation, or government is fully aware, let alone in charge, of these systems and their consequences.[10]

It's a daunting thought, and it would be difficult not to follow it up with an obvious question: What could possibly go wrong?

CRACKS IN THE FOUNDATIONS

We've all seen the stories in the news, or breathless first-person accounts on social media, about internet-connected devices doing something best described as "creepy." Maybe you read the one about the guy who was talking privately to his doctor about a sexual health problem, and then the next time he opened his laptop, all of the ads were for erectile dysfunction pills. Maybe you heard about the parents who walked into their child's bedroom in the middle of the night to discover that a stranger's husky voice was whispering through the baby monitor to their sleeping infant. Maybe you saw a video of someone's internet-connected vacuum cleaner "acting drunk"—spinning around and bumping into furniture like someone who's had one too many, knocking over lamps and wreaking havoc in the home. There are thousands of stories like this, self-reported by users of Instagram, TikTok, and Reddit, occasionally picked up and verified by journalistic outlets, and on rare occasions acknowledged and "fixed" by the companies responsible for building and selling the devices in question.

Whether each of these tales is individually true or factually accurate is almost irrelevant at this point, because collectively they point to a widespread, almost folkloric awareness that we have surrounded ourselves with technology that is capricious, unaccountable, and potentially hazardous. As with classic folktales about the pitfalls of living among ghosts, djinns, and fairies, most people's reactions tend to be along the lines of "These things are beyond my understanding, and there's nothing I can do about it; I just hope it doesn't happen to me." The difference, of course, is that, in the case of internet-connected devices, we can, or at least *should*, have some control over the role they play in our lives and a fair degree of confidence in their ability to improve our standard of living. It's important to remember that hacks, bugs, and other cases of cyberphysical systems going terribly wrong don't just *happen*. They are the results of mistakes, oversights, cost cutting, negligence, poor regulation, calculated risk, and sometimes even malice on the part of people who build these systems and the technologies they rely on. They may also be the inevitable outcome of building a system so vast that it can't possibly be understood in its totality.

There are many factors undermining the security and reliability of cyberphysical systems. One of the simplest is that, despite building futuristic technology, technologists don't often think much about the future. In 2020, cybersecurity firm Forescout issued a research report detailing what it considered critical vulnerabilities in millions of IoT devices manufactured by hundreds of companies. According to Forescout's analysis, these vulnerabilities would "allow an attacker to take control of a device," giving them access to the data being collected and stored and letting them use these devices "as part of large attack campaigns, such as botnets, without [the owners] being aware."[11]

One might assume that, once identified, these vulnerabilities could be fixed by the companies that sold the vulnerable devices, or at least patched by the people who bought them. Unfortunately, even had they known or cared about the problem, manufacturers and consumers alike were powerless to do anything about it. That's because the vulnerabilities are part of an essential building block of computer networking called the TCP/IP stack that nobody owns, but everybody uses. Because this foundational code, which helps individual devices connect to digital networks, isn't the responsibility (or a profitable asset) of any individual or company, nobody has the power or the incentive to do the hard work of fixing it.[12] As a result, the threat is here to stay—at least until the millions of compromised devices are unplugged and turned into e-waste.

Forescout called this particular batch of vulnerabilities AMNESIA:33, because they affect computer memory. However, the real problem in this case wasn't difficulty remembering the past, but an inability to consider the future. When technology developers cut corners by integrating bits of unaccountable and unvetted code into their own products, they pass on the risks associated with that buggy software to all of the customers who use their products. Even worse, as a vulnerable building block like a TCP/IP stack is adopted more widely, the flaws it contains become harder and harder to eradicate, because so many other pieces of technology are dependent on it. It's the difference between replacing a broken brick at the bottom of a pyramid and replacing a brick at its summit.

The AMNESIA:33 vulnerabilities are only one example of a much more widespread phenomenon.[13] Not only are many of the building blocks used by technologists flawed, but others are intentionally made vulnerable, because many governments around the world (including that of the United States) have demanded that technologists engineer so-called back doors into their code, allowing eavesdroppers to remotely control and surveil networked devices without the knowledge or consent of their users. While these policies are often enacted in the name of fighting terrorism or other serious social problems, the back doors are inevitably discovered and exploited by others with less noble purposes, such as identity theft, corporate espionage, and cyberstalking.

To make matters even worse, when cybersecurity researchers discover these vulnerabilities, or speak out against them, they risk being prosecuted as hackers under laws like the United States' Computer Fraud and Abuse Act (CFAA), or fired from their jobs for undermining their employers' public images and bottom lines. Additionally, many researchers end up selling their discoveries secretly to the highest bidder (such as a national intelligence bureau) rather than reporting them altruistically to the manufacturer and the public at large, which means that critical vulnerabilities can remain invisible to the companies that make networked devices—and, by extension, their users.

All of these factors combined have helped to make the IoT like a city of pyramids built on a foundation of broken bricks, threatening to expose its residents to the elements, or worse, to crumble and fall. And these vulnerabilities make it shockingly easy for schemers and scammers to peek through the cracks, snooping on private lives and businesses with ease. According to one recent study by a multinational team of computer science professors, 77 percent of all IoT devices connected to the internet can be physically located and individually identified *just by sampling and analyzing traffic on the open internet*—no hacking necessary.[14] It also means that each of us (yes, even you!) is probably leaking all kinds of very personal information to unaccountable third parties every single day as we go about our professional and personal lives.

If the IoT is such a security disaster and so prone to vulnerabilities and hacks that it has rapidly developed its own genre of creepy cautionary tales, why would anyone buy or sell these devices? For consumers, the answer is simple: they're cheap, convenient, and sometimes even kind of cool. Besides, there aren't a lot of alternatives out there—when did you last see a plain old "dumb" television for sale on Black Friday? For tech companies, the answer is even simpler: there's a lot of money in it.

THE INTERNET OF YOU

Let's be clear about this: smart devices are, by definition and by design, surveillance devices. Amazon may or may not make a small profit selling you a smart speaker to place in your living room for $29.99, but from the company's perspective, the real economic benefit comes after you've opened the box and plugged in the device. Not only does the voice-activated Alexa search engine make it easier for customers to buy more products and stream more content directly from Amazon, but the device drastically increases the amount of personal data the company can collect and analyze, including (according to a recent class action lawsuit) voice recordings captured without the knowledge or consent of the people being recorded.[15]

We don't mean to pick on Amazon. Smart appliances from Google, Apple, and Meta do pretty much the same thing for the same reasons, as do countless other similar products from companies large and small around the globe. Many consumers may consider the loss of privacy a worthwhile trade-off for convenience. If you stick an always-on, internet-connected microphone in the middle of your house, you probably shouldn't be too surprised to find out it's eavesdropping on your conversations while it awaits your next command. And if telling your music preferences to a corporate database somewhere means you get to spend $30 on a voice-activated sound system instead of $300 or $3,000, that's another big incentive not to read the fine print of the privacy policy in too much detail. But while savvier consumers may have made their peace with the privacy limitations of their smart speakers, very few people are fully aware—let alone comfortable—with the range of devices, the variety and volume of data, and the number of interested parties involved in surveilling their daily lives via other cyber-physical systems.

First of all, nearly everything that *can* be connected to the internet *is* being connected to the internet—including banal and intimate items like forks, sex toys, and toilets. In almost every case, those networked devices are collecting and monetizing some form of data about their users in order to profile us, advertise to us, and predict or influence our future behavior. And it's not only our own devices that collect these data; when we enter a friend's home, a place of business, or even a public space, we have no way of knowing the number and range of cyberphysical systems taking note of our presence, recording our activities, and analyzing the data we produce. Like us, you may even have gotten into the habit of saying, "Alexa, play 'Despacito'" when you walk into a stranger's home or a hotel room, just to see whether a voice-activated speaker is listening.

Voice-activated appliances and voice recordings are just the tip of the iceberg. Increasingly, these devices don't just capture audio or video—they aim to map three-dimensional space, ideally in real time. Robotic vacuum cleaners like Roomba (which is now owned by Amazon) currently use a technology called simultaneous localization and mapping to create a detailed map of your home as they clean, which is then published to a "home knowledge cloud" and sold off to third parties who would like access to it. Companies like Amazon and Google have also begun developing apps around new, ultra-sensitive radar sensors, which will not only allow users to interact with their devices through physical gestures, but will collect highly detailed, ongoing information about users' bodies and the places they inhabit. Amazon has even sought regulatory approval to use radars like this to provide sleep-tracking services to users, promising, like Sting, to keep track of "every breath you take."

An analogous technology called light detection and ranging (LiDAR) uses laser imaging to create precise three-dimensional maps of objects and spaces. Originally, LiDAR was used in large-scale, outdoor applications, such as autonomous vehicle navigation and monitoring crop yield for big agriculture; in recent years, however, as the technology has become cheaper and more compact, it has been integrated into more IoT appliances and personal devices. In fact, every iPhone "pro" model manufactured since 2020 includes a LiDAR sensor, and users are being encouraged by app developers to capture three-dimensional models of the people, places, and things in their lives, and to upload them to online databases.

Not all smart devices aim to surveil human bodies and physical space. A rapidly developing technology ecosystem allows cyberphysical systems to track one another and to share information in their own device-to-device networks. Some products were designed to do this from the start, such as Tiles and Apple's competing product AirTags (the little, plastic, Bluetooth-enabled beacons you can put in your wallet or on your keychain so they'll alert your phone if they get lost). But, increasingly, the devices in our homes and environments seek out connections to one another without bothering to let us know.

For instance, a 2019 upgrade to Apple's iOS mobile operating system revealed to users that Facebook had been using iPhones' Bluetooth antennae for years to surreptitiously track their owners' locations and interactions with other Facebook users in the real world—even if their phones were in their pockets and the Facebook app was closed.[16] And, more recently, technology researchers have reported on an entire "beacosystem" (a portmanteau of "beacon" and "ecosystem") that tracks and identifies people's personal devices as they travel through public and semipublic spaces such as retailers, airports, and music festivals.[17] Even the fancy new digital billboards and kiosks that now crowd the corners of many metropolitan centers frequently have Bluetooth beacons embedded in them. And lest you think you can prevent these unsolicited connections by turning off the Bluetooth and Wi-Fi signals on your personal devices, there are other, even more surreptitious systems for communication between devices, such as ultrasound device tracking, a technology commonly used in Android-based mobile apps in which beacons use high-pitched audio outside the range of human hearing to share data with your phone, smartwatch, or other smart appliances via their speakers and microphones.

We spoke to Vasilios Mavroudis, a computer security researcher at University College London, about the troubling implications of ultrasound tracking and other data conduits that, in his words, "move between the cracks of two systems." Part of the problem, he told us, is that technologists outside of the security field have a hard time imagining the creative ways in which hackers might sidestep the security protocols they build into their platforms and devices. As he put it, "We've done a not particularly good job of trying to incorporate side channels into threat models." This is to be expected: as any good hacker will tell you, the key to success isn't only in figuring out the gaps and limitations in a given piece of tech, but also the weaknesses of the people who build and use it.

However, Mavroudis reminded us that not all data security threats come from without; many of them are built into technologies intentionally, for reasons like those discussed above. This pervasive surveillance-by-design ethic makes it especially difficult for end users to opt out of unwanted tracking, because it violates the very purpose for which the tech was built. Consequently, as Mavroudis explained, there's a relationship between a system's usability and its vulnerability to data harvesting. Often, the tracking capabilities of a technology are a feature, not a bug—they're central to enabling its core functions.

This raises the question of whether a truly secure communication technology platform can ever be built. As Mavroudis put it, "You can leave everything down to the bare basics, but then you're going to have zero uses." Anyone who's ever tried to surf the web with an ad blocker installed in their browser has experienced some version of this truism—sites don't load fully, images don't appear, links can't be clicked. To paraphrase the WOPR computer in the classic 1983 film *WarGames*, when the entire system is built to extract your data, the only winning move is not to log on.[18]

And yet, as Laura DeNardis reminded us, the rise of cyberphysical systems means there's no longer any off switch for the internet or the data harvesting it enables. And that means that data collection has now become an integral part of virtually everything we do, individually and collectively. Tech gurus have been telling us for years that "data is the new oil," meaning that it's a hot new extractive commodity fueling an industrial boom. But the metaphor can be extended even further: just as petroleum became a building block in scores of secondary industries, such as plastics, automobiles, and beauty products, limitless data is now a given for inventors and entrepreneurs dreaming up new business opportunities across many different fields. And, as with petrochemicals, these new uses haven't merely been discovered; they've been promoted by the tech industry itself as it seeks to build secondary profits and a longer life for data, its core resource, while making information products essential to every social institution. So even if there were an off switch, nobody would dare to use it.

BIG DATA IS BIG BUSINESS

Once our data have been harvested from smart devices, thermostats, voice-activated speakers, billboards, and baby monitors; they are mined for relevant bits of information, reassembled into valuable commodities, and sold on the open market. Though consumer-facing tech companies like Amazon and Google often use such data to enhance their own services (e.g., encouraging us to buy more stuff, or writing our emails using predictive algorithms), they also participate in a larger marketplace in which individual profiles and aggregate consumer data are collected and sold by companies you've probably never heard of, often in combination with proprietary analytical tools. These companies perform a complex array of tasks and offer a broad range of services, but they're generally referred to with the simple umbrella term *data brokers.*[19]

Some of the most sensationalistic uses of brokered data take place in real time. For instance, in 2018, Burger King launched a promotion that gave potential customers a coupon for a one-cent Whopper hamburger if they were within six hundred feet of a franchise of the company's chief competitor, McDonald's. The ability to pinpoint customers and competitors geographically is something that many of us in the 2020s probably take for granted. But it's important to understand how recent this development is, and how much it depends on the existence of an entire cyberphysical architecture that includes GPS satellites, a free and open internet, and—perhaps most importantly—billions of internet-connected smartphones in billions of pockets around the world. While we used to speak about the internet in its early days as *cyberspace*, using the metaphor of an information network modeled on the physical world, in many ways its impact has been the reverse: the physical world has come to behave more and more like the internet.

Despite the flashy appeal of a real-time, GPS-targeted coupon, most uses of brokered data are cumulative and operate on a longer timescale, such as an insurance company deciding what premium to charge you based on data points that may come from several decades of activity. In fact, the entire data broker industry is built on the premise that the value of data collected from a consumer today may not be fully apparent until many more years have elapsed. We spoke to Dennis Crowley, the founder and former CEO of Foursquare, a company that collects and sells geographic location–based data about pedestrian traffic harvested from smartphones, or what he refers to as "a cookie for the real world."[20] As he explained to us, it's hard to know whether an individual data point, or even a given dataset, will ever bear fruit financially. That's why data brokers typically collect far more information than they can use, hedging their bets that it may turn out to be valuable in the long run. In Crowley's words, "Sometimes, it's easier just to store more stuff than it is to take the risk of deleting stuff." Therefore, it's no coincidence that the data broker industry has bloomed as the prices for data storage have imploded.[21]

Data brokerage is a big business. By some credible estimates, mobile phone location history data in the United States *alone* were worth about $12 billion in 2020 (roughly the same amount as the entire American recorded music industry).[22] Nobody knows the collective value of all the data brokered on a global scale each year, but conservative estimates by business intelligence firms place the legal market somewhere around $300–500 billion annually, while a shadow black market for illegally collected consumer data also reaps untold billions on the dark web and elsewhere.

Even legal data collection isn't necessarily ethical, however, because data privacy laws in the United States and elsewhere have a long way to go before they catch up to basic, widespread expectations about what kinds of surveillance are fair or reasonable and how personal data should be handled by state and commercial institutions. And while companies like Foursquare were built on an opt-in model, allowing consumers to tell businesses when they've walked in the door, and taking pains to shield individual consumers from identification without their permission (though even that is hardly a guarantee of personal data security, as we have discussed), they are, unfortunately, the exception rather than the rule. "A lot of companies out there are just sneaking whatever they can into whatever apps they can" as data harvesting tools, Crowley told us. "A lot of shitty poker games and solitaire games and flashlight apps and Tamagotchi stuff was just designed to collect location data and sell it to third-party brokers." Crowley's analysis is spot-on; as we were writing this chapter, tech journalism site *The Markup* broke the news that Life360, a massively popular location-tracking app marketed as a tool for parents to keep track of their kids, has been selling children's location data to about a dozen data brokers, making it one of the largest sources of such data in the marketplace.[23]

Even though most of these brokerages operate outside the public eye and charge significant fees for access to their data, it's possible for everyday internet users to get a glimpse into their operations without having specialized expertise or spending any money. We signed up for free access to location-based data from a broker called Near, which claims it has detailed data about 1.6 billion people in forty-four countries.[24] The tool we used (called Vista) allows you to choose a specific location from a map, such as a business or place of residence, and, in about fifteen minutes, get a detailed profile of all the people in their database who have visited that spot in the previous year.

First, we selected one of our own homes. We were able to see clearly the common daytime and evening locations associated with individuals who visited us during 2021, including the family home of a teaching assistant, the homes of several of our children's friends, and, presumably, the homes of our mail carrier, food delivery workers, and other service workers who came to our house during the COVID-19 pandemic. We were also able to establish that the most common age of visitor to our home was under eighteen (accurate, as we have two school-aged children), that 50 percent of our visitors were white, and that the median household income of our visitors was $96,487.

Next, we selected a site with a higher degree of public interest: Donald Trump's Mar-a-Lago resort in Palm Beach, Florida, which is frequently cited as a base of his political operations during and after his presidency. Using Vista, we were able to learn a lot about the people who visited the club between December 2020 and December 2021. Predictably, the demographic profiles of its visitors were different from ours; about two-thirds of Mar-a-Lago visitors were white, and the most common age group was people over sixty-five years old. The median household income of visitors was $84,352, perhaps reflecting the relatively low wages of the numerous workers at the club.

Even more interesting were the common evening and daytime locations of visitors to Mar-a-Lago. While most were American (with over 70 percent coming from Florida and none from Wyoming, North Dakota, or Vermont), visitors arrived from every populated continent. We were able to pinpoint on a map the specific locations associated with the devices belonging to individual international visitors to the club, including

- a private residence in Akure, Nigeria, presumably linked to Abraham O. Adeyemi, the Akure-based pastor whose video of a pro-Trump parade was retweeted by Trump on November 3, 2020;

- a private home in the ritzy gated golfing community of Nordelta, on the northern outskirts of Buenos Aires;
- a business address on the outskirts of Moscow registered as the office of Agro Allied Agency, a petrochemical company that produces fertilizer and diesel gasoline;
- a residential and business complex in Kherson, a port city in southern Ukraine; and
- a luxury residence on a cul-de-sac in the upscale Talbiya neighborhood of Jerusalem.

We are not investigative journalists, and we have neither the time nor the expertise to look more deeply into these and other data points associated with visitors to Trump's Mar-a-Lago club, or any other location. Our aim here is simply to demonstrate how easily the data collected from cyberphysical systems and commodified by data brokers can be used in ways that have little or nothing to do with the targeted advertising and business intelligence applications promoted by the brokers themselves. In a matter of minutes, we were able to find the possible overseas addresses of a US president's associates using a free version of an online tool from a single broker. So imagine what kinds of information may be developed by an experienced data analyst or investigator using the premium versions of several of these tools in conjunction with one another, or by a resourceful hacker collecting data directly from our devices.

POLICE AND THIEVES

Even though most of the collection, analysis, and sale of data harvested from cyberphysical systems are done for the purposes of targeted advertising, there are several other interested parties who routinely use these data, as well, for very different purposes. These include large-scale employers, law enforcement agencies, and criminals.

Some of the employers who have been most aggressive in their use of data collection and analysis to monitor employee behavior are the same companies that build and maintain large-scale surveillance platforms. For instance, most American consumers probably think of Amazon as an online superstore, and maybe secondarily as a streaming video company, a smart device manufacturer, or the parent company of the Whole Foods supermarket chain. But about one-eighth of its revenues (currently $40–50 billion annually, out of a total of around $400 billion) come from Amazon Web Services, a suite of cloud-based data and infrastructure offerings that includes, among other things, surveillance platforms like Rekognition, an AI-based facial recognition service that the company licenses to both government agencies and commercial clientele. Similarly, while many people might still think of IBM as a manufacturer of computer hardware and software, it has evolved in recent years into a data and computing services company, and its AI division now accounts for about one-third of its annual revenues.

As it turns out, Amazon and IBM have used their considerable data surveillance and analysis powers to keep tabs on their own employees, as well. Amazon installed devices in its delivery vans that use AI-enabled video cameras and biometric sensors to keep constant track of their drivers, sending reports to their dispatchers if they yawn, seem distracted, or look at their phones. Similarly, the company installed a system in its warehouses that tracks every employee and scores their productivity, taking points off if a worker appears to be taking too many breaks or moving too slowly. As news outlets like *The Verge* and *Business Insider* have reported, these systems have been used to terminate or reduce compensation for thousands of employees with little to no human oversight, let alone a reasonable appeals process.[25]

While Amazon's techniques police the physical behaviors of poorly paid manual laborers, IBM has developed its own AI system to track the intellectual and emotional states of its higher-paid white collar workers. IBM claims that by analyzing employees' task completion rates and educational development, in combination with what former CEO Ginni Rometty called a "secret sauce" of other data and analytics, it can predict with 95 percent accuracy which valuable employees are in danger of leaving for another job (in which case efforts are made to retain them), and which workers are underperforming and should therefore be terminated or "reskilled" (given a different job).[26] Neither Amazon nor IBM has denied surveilling its workers in order to predict, police, and punish their behaviors; to the contrary, both companies have publicly touted the effectiveness of these AI-assisted human resources policies. IBM is even using its own self-reported success as a proof of concept to sell the service to other white-collar employers.

Over the past decade or so, law enforcement agencies both locally and nationally have become some of the biggest customers for device-based surveillance, data brokers, and predictive analytics. Though constitutional protections in countries like the United States limit the scope of direct surveillance by police departments and federal agencies, there are far fewer barriers preventing these agencies from buying commercial data about consumers who have opted into the terms of service attached to their personal devices and apps, or from using public surveillance platforms that don't technically invade the private lives of their targets.[27]

The examples of law enforcement accessing and acting on data harvested from cyberphysical networks are too numerous to recount. In some cases, data brokers have provided bulk information to law enforcement and other government agencies without either a court order or a clear sense of how it might be used. For instance, Amazon's widely adopted, video-equipped Ring doorbell product now serves as a de facto public CCTV system with millions of live cameras recording street traffic in thousands of neighborhoods throughout the United States. The company provides bulk video footage, along with analytical tools, to about one-eighth of all police departments in the country (over 2,100 in total), as well as more than 450 fire departments, and has handed over video related to individual customers without a warrant or consent on numerous occasions.[28] Along similar lines, data broker Veraset recently provided in-depth GPS data for hundreds of thousands of individual phones and devices in the Washington, DC, region to the DC government, ostensibly to track potential COVID-19 infections, but reportedly without any checks or balances to prevent other uses and abuses of the data.[29]

Law enforcement increasingly targets individuals using these commercial surveillance methods in order to sidestep the legal or logistical obstacles that are supposed to rein them in. For instance, the US Secret Service reportedly bought location data for individual suspects from a data broker called Babel Street while investigating a financial fraud operation.[30] Along similar lines, the FBI reportedly used a bevy of commercial data sources, some of them without a warrant, to track down the identities and actions of rioters who participated in the attempted coup at the US capitol building on January 6, 2021.[31] Data collected from smart devices and other cyberphysical systems have also begun to appear more frequently in court cases. In one recent, high-profile case, a man in his nineties was tried for the brutal murder of his stepdaughter, based largely on evidence from her Fitbit wearable device and video footage from Amazon Ring cameras.[32] In another case, a Florida man was exonerated of the charge of murdering his girlfriend after police filed a warrant for audio footage from her Amazon Echo smart speaker and found no evidence of him participating in the killing.[33]

Are these uses of data harvested from smart devices and IoT networks a good thing or a bad thing? It depends on your perspective. On the one hand, most people would probably agree that it's important to hold murderers and insurrectionists accountable for their violent crimes, and that everybody benefits if businesses can incentivize their employees not to quit. On the other hand, police sidestepping judicial warrants by buying data from a broker could be the beginning of a slippery slope to undermining constitutional rights. And who would ever want to be spied on by their boss, especially if it meant they might lose their pay or their job as a result of doing something normal and involuntary, like yawning or seeming distracted?

Yet the same features that make cyberphysical devices so valuable to consumers, employers, and law enforcement agencies also make them a prime target for criminals—and this is a problem everyone (except cybercriminals) can agree on. As we discussed above, the IoT is fundamentally insecure; as one recent study by cybersecurity firm Palo Alto Networks showed, 98 percent of such devices share unencrypted traffic over the public internet, meaning that anyone with the right tools can access to the data they collect, and more than half of these devices are at risk of "medium- or high-severity attacks."[34]

Cybercrime has many faces. In the simplest case, a hacker might intercept personal data from a user's device and use it to open a credit card or bank account in their name. Others might install malware on these devices and use that malware to stage larger online attacks, or to mine cryptocurrencies like Bitcoin without the knowledge or permission of the devices' owners. A more enterprising hacker might use sensitive information from IoT devices, such as video footage or health data, to blackmail or stalk a victim. Sometimes, it's not an anonymous hacker abusing data, but an intimate friend or enemy; as Laura DeNardis told us, "Every law enforcement officer knows that this is a huge problem in domestic abuse cases."

Often, cybercriminals target not an individual, but an entire institution. IoT vulnerabilities have already contributed to hacked power grids, supply chain disruptions, and travel shutdowns around the globe. At the time of writing, one of the authors lives in a state that has been experiencing a ransomware attack for several months: hackers have entered the Maryland Department of Health computer network and locked officials out of their own COVID-19 reporting data unless they pay a hefty ransom fee, making it far more difficult for health workers and everyday citizens to confront the deadly pandemic, potentially leading to more hospitalizations and deaths.[35]

The world was never a particularly safe place; stalkers and abusers terrorized their targets, and wars and pandemics decimated populations long before mathematics were invented, let alone digital databases. But the secret life of data has helped to create a new architecture for society that is both more complex and less visible than what came before. To use the language of security researchers, the *threat surface*—the area vulnerable to attack— has become much larger and harder to protect. In the former Eastern Bloc nations, secret police notoriously tapped phones and tailed pedestrians—but each spy could listen to only so many hours of conversation and tail only one person at a time. Today, with minimal expenditure, expertise, and effort, any police officer, advertiser, spurned lover, or potential blackmailer can collect far more detailed information about an individual or an entire city. We haven't yet adjusted our norms and ethics, let alone our laws and technologies, to accommodate this new reality.

Of course, as we mentioned at the outset of this chapter, there are many benefits to cyberphysical systems, IoT networks, and data brokers. They can provide convenience, boost efficiency, and sometimes even aid in the pursuit of justice. And, to be honest, talking robots are still pretty cool. But we must understand that when we talk back to them, it's not merely our own digital assistants who are listening, but an entire network of anonymous and unaccountable actors, in the present day and far beyond the foreseeable future, analyzing our data in order to profile our personalities, predict our behaviors, and influence our decisions. In the final analysis, we must ask: is it worth it? Ultimately, it depends on where you are in this vast web of data, and what you stand to gain or lose.

5 THE SECRET DATA OF LIFE

○

On July 17, 1537, a thirty-nine-year-old woman named Janet Douglas was tied to a stake and burned alive on Castle Hill, in Edinburgh, Scotland. Also known as Lady Glamis, she was accused by King James V (her brother's stepson) of using black magic to plot against his throne. Sadly, she was one of many women to meet such a fate; by some accounts, roughly two thousand accused witches were burned to death on Castle Hill between the fifteenth and eighteenth centuries. Nearly half a millennium later, a woman's life was saved on the very same spot. In this case, she survived thanks to a technological tool masquerading as magic.

Bal Gill was a forty-one-year-old tourist from the London area, visiting Scotland with her family in 2019. After trekking around Edinburgh Castle, they decided to visit Camera Obscura and World of Illusions, a nearby museum dedicated to showcasing the work of "artists, inventors and technical wizards," with exhibits named Bewilderworld, Light Fantastic, and the Magic Gallery.[1]

When Bal and her family reached the museum's third floor, they were intrigued by an exhibit that films visitors with a thermal camera and projects heat maps of their bodies onto a giant screen. As Gill later wrote, she and her family started waving their arms and goofing off in front of the camera, watching their blue, green, yellow, and red-tinted thermal images echoed on the wall. Gill thought she noticed a red patch, indicating higher temperature, on her own left breast. Nobody else in her family seemed to have a similar discoloration.

After returning home, Gill looked through the photos she'd taken during the trip. The red patch over her breast showed up clearly in one image, confirming her memory, so she started Googling to see what it meant. She discovered that thermal imaging technology is sometimes used to detect breast cancer, so she made an appointment with her physician, and sure enough was diagnosed with the disease. Thankfully, it was in its early stages, and Gill was able to treat her cancer successfully. After two surgeries, she wrote a letter thanking the museum for her "life-changing visit." Gill said she owed her survival to the unintentional medical imaging. "Without that camera," she wrote, "I would never have known."[2]

Thermal imaging technology like the camera that caught Gill's tumor is an example of a rapidly growing and changing set of technologies that produce data about living bodies. Collectively, these technologies are typically referred to as *biometrics* (meaning "life-measuring"), and they run the gamut from facial recognition algorithms to wearable heart rate monitors to the fingerprint scanners on our phones and laptops. In recent years, as biometric technologies have become smaller, cheaper, and more sophisticated, they have found their way into many of the systems and devices we interact with on a daily basis, and thus have become thoroughly integrated into both public spaces and our private lives.

The companies that collect biometric data store and analyze them in digital formats and on networks that were developed for uses like advertising, law enforcement, and health care. More often than not, that means these data are integrated with other kinds of information to develop complex portraits of us as individuals and populations. Those portraits are then used by a variety of commercial interests and governmental bodies in ways that combine biological and social factors to make important decisions about our lives—often without our full knowledge, let alone our informed consent. Once a fingerprint, a face, or a strand of DNA is coded into ones and zeroes and stored in a database, there's no telling who might end up in possession of it, or what uses they might find for it. In other words, biometric data, once the exclusive domain of individuals and their doctors, now has a secret life of its own.

Sometimes, these unexpected consequences can verge on the miraculous, saving lives like Gill's. But far too often, they end up being used in ways that can seem more like a modern-day witch hunt, condemning people to horrible fates based on spurious reasoning and incomplete information without transparency or recourse. While nobody is likely to be burned at the stake, a shocking number of people may be denied medical coverage, bail, or basic human rights based on some third party's analysis of their biometric data. Therefore, it's essential that we understand what these technologies do, how they might be used, and what kinds of larger social consequences biometrics might lead to before we invite them any deeper into our lives.

BIOMETRICS FOR THE WIN

Bal Gill's run-in with a thermal imaging camera was a happy accident, a case of biometric technology intended for public entertainment turning out to have a private medical use. And, to be sure, millions of people deliberately use biometrics, from wearable fitness trackers to blood glucose meters to temperature scanners at airports and event spaces, for public and private health benefits. But, increasingly, biometrics are also being used for purposes that extend far beyond the spheres of health and medicine. Many of these uses have clear social benefits. Other cases are more tenuous, benefiting some people at a cost to others. And some are just plain dystopian.

Let's start with the good stuff. One of the newer tricks in the biometric bag is called eye tracking. You've probably heard of it, or even seen it in action. True to its name, this technology uses a combination of video cameras and proprietary algorithms to analyze where a person is looking, creating a quantifiable measure of attention that may be used for optimizing advertisements, improving software interface designs, or even profiling an individual's tastes and predilections. But a recent initiative applies this biometric tool to a different use: making museums more appealing to art lovers.

This project, called ShareArt, is the brainchild of Italy's National Agency for New Technologies, Energy and Sustainable Economic Development (ENEA), a government research division that develops technologies for a range of purposes from nuclear safety to cultural heritage. Because tourism is such an important element of Italy's economy, optimizing museums to suit the tastes of their visitors is a matter of national consequence. This is where eye tracking comes in. In 2021, when Italian museums reopened following closures related to the COVID-19 pandemic, the Instituzione Bologna Musei installed the first fourteen ShareArt devices, using cameras mounted alongside paintings and sculptures to collect data about which artworks, and which features of these artworks, commanded the most attention from museum visitors.[3] At the time of writing, the results are being analyzed by data scientists at ENEA to develop recommendations for the museum curators that could be used to increase attendance and, ultimately, to help grow Italy's tourism revenues.[4]

Another example of biometric technology deployed in the public interest is the use of DNA analysis and other forensic techniques to reconstruct historical events that were either ignored or intended to be forgotten. For instance, in 2021, researchers at the Smithsonian Institution shared the results of an eight-year study that examined the remains of twenty-nine enslaved people who had been buried more than two centuries earlier near a Maryland iron forge called the Catoctin Furnace. When their bones were first exhumed to make way for a road in the 1980s, "nobody even cared" who they were, as Sharon Burnston, the archeologist who directed the exhumation, told the *Washington Post*.[5] However, in the past forty years, both genetic forensic techniques and the public reckoning with America's history of slavery have advanced to the point where new information can be gained, and a new process of national self-evaluation can unfold, from reexamining these relics. A baby buried with his enslaved African American mother was revealed to have a white father. There was widespread evidence of physical abuse and degrading work conditions, ranging from zinc poisoning to severe labor-related deformities. Five or six discrete family groups were reconstructed, based on similarities in the DNA extracted from the bones.

At a time when the United States is at war with itself about whether and how the legacy of Atlantic slavery should be taught in public schools and universities, and when battles over civil rights, labor relations, and public health continue to dominate news cycles, these kinds of discoveries are not mere

Figure 5.1
Catoctin Furnace in the present day. *Source:* Photo by Aram Sinnreich.

historical curiosities. The secret life of biometric data is becoming an indispensable tool in public debate and civic engagement and a valuable means of validation for inconvenient truths, both historical and contemporary. Yet, unfortunately, biometrics are being used for other political purposes, as well—ones that threaten to undermine democratic principles and escalate social tensions, with potentially devastating consequences.

FACIAL RECOGNITION AT A GLANCE

Before we go any further, we should spend a moment discussing one of the most prevalent and hotly contested forms of biometric data collection: facial recognition technology (FRT). This term covers a wide range of different tools and applications, but at its core, it comes down to one basic function: using computer algorithms, typically AI, to establish the identity of an individual whose face has been recorded by a camera or other sensor.

FRT is a great example of a technology that has become ubiquitous because of a perfect storm of several other interdependent technological and social developments, including rapid improvement and plummeting prices of digital cameras; the proliferation of the mobile internet; near-universal adoption of selfie-saturated social media platforms throughout industrialized societies; the rise of data brokers and surveillance capitalism; the escalating level of state surveillance in public spaces; geometric growth in the power of machine learning algorithms; and the coalescence of all these trends in the form of the smartphone. In fact, as you probably know, most modern smartphones now allow users to unlock them using FRT instead of a passcode or a fingerprint scan.

Like other forms of machine learning, FRT is created by assembling a large database of seed images and performing feature extraction on the underlying data to identify common attributes among those images. For instance, the widely used CelebA dataset, first published by computer scientists at the Chinese University of Hong Kong in 2015, contains over 200,000 images of more than ten thousand different celebrities' faces, each with forty discretely coded attributes (such as "wearing hat," "mustache," and "pointy nose").

These attributes can be combined with the measurable dimensions of a face, such as the distance between its eyes or the length of an upper lip, to create a *faceprint* for a given individual—a quantifiable collection of details that add up to a unique identity, easily reidentified when a new image is analyzed by the algorithm. That's why the CelebA dataset isn't useful only for figuring out where you've seen that guy on that show before—its principal function is as a training set enabling an FRT model to identify anyone's face, even those of us noncelebrities. Once you've taught a computer to recognize a few thousand pointy noses, it can theoretically recognize millions more with relative ease and accuracy.

Using celebrity photos to train FRT algorithms didn't just clear the technological path for widespread surveillance; it also cleared the way ethically. We're accustomed to celebrities being in the public eye, and images of their faces are ubiquitous on magazine covers, in makeup ads, and in memes. How many times have you seen Mandy Patinkin's charming smile appear on a social media feed with the phrase "You keep using that word. I do not think it means what you think it means" emblazoned across it in white block letters? Even legally speaking, in many countries, celebrities have more limited privacy rights than other people enjoy—that's part of the reason there's a thriving and (mostly) legal market for paparazzi photos.

But, as with so many issues surrounding data and society, it's a slippery slope. In 2021, the CEO of Clearview AI, which sells FRT tools to thousands of law enforcement and government agencies, announced that the company had amassed a database of more than ten billion images of everyday, not famous internet users by scraping their photos from publicly available websites like Facebook, Instagram, and Twitter.[6] The announcement sparked a fair amount of handwringing, and the company has faced some repercussions, including a lawsuit by the American Civil Liberties Union and an order from the Australian government to expunge its citizens' biometric data. But the public outcry was limited, and the story probably garnered far less attention than it would have if datasets like CelebA hadn't already normalized unpermissioned collection and analysis of facial images for the better part of a decade. At the time of writing, the company is apparently thriving, and *Time* magazine named it one of the one hundred most influential companies of 2021.[7]

PRIVATE EYES, THEY'RE WATCHING YOU

One of the principal uses of biometric technology beyond the world of medicine and health care is surveillance. As new technologies and business practices have gotten better and better at turning the complex internal workings of the human body into strings of ones and zeroes, most of the markets and industries built to monetize mouse clicks and track devices have been adapted to accommodate the resulting torrent of biometric data. As in the case of Clearview AI's facial recognition technology, many of the most enthusiastic buyers are in law enforcement and similar areas, but the range of interested parties is much wider, encompassing diverse state and commercial institutions, as well as rogue actors like criminals and terrorists.

As with more traditional forms of surveillance, biometrics are not deployed randomly, neutrally, or indiscriminately; they're frequently used in ways that reinforce or more deeply institutionalize existing social inequities and power structures. In Israel, for instance, the military has been using an FRT called Blue Wolf to surveil Palestinian civilians in the occupied West Bank. As the *Washington Post* reported, in order to build a database of faces, soldiers were enlisted in a competition to see who could photograph the greatest number of Palestinians, with prizes awarded to the most prolific units. Once the database was assembled, every Israeli soldier in the region was given a special smartphone app that allowed them to scan the faces of pedestrians and immediately receive a color-coded alert instructing them to detain, arrest, or leave be the Palestinian whose photo they'd taken. FRT has also been integrated into cameras throughout the divided city of Hebron, monitoring the Palestinian population in real time and even peeping into some private homes.[8]

A similarly nightmarish version of biometric surveillance is currently underway in China, where the government uses a combination of genetic profiling and FRT to control the population of Uighurs, a predominantly Muslim ethnic minority of roughly eleven million people. According to a report by the US Congressional Research Service, more than one million Uighurs have been detained in brutal "reeducation centers" and subjected to abuses including forced labor, sexual assault, and torture by the Chinese government since 2017.[9]

As investigative journalists at outlets such as the *New York Times* have reported in recent years, China is actively pursuing a biometric surveillance strategy that will allow the government to keep tabs on the other ten million Uighurs in the general population.[10] The strategy involves at least two different biometric data sources. First, with the help of American researchers and firms, the Chinese government has been developing the world's largest DNA database, profiling both Chinese nationals and other people from around the world. This initiative has included the coercive collection of DNA samples from countless Uighur civilians and detainees with the explicit aim of creating a *DNA phenotype*—a physical profile of the genetic information shared by an ethnic group—for Uighurs. Using this profile in conjunction with its widely deployed FRT and robust databases of existing faces, China reportedly aims to create a "Uighur alarm" that will automatically scan public video footage and alert authorities when a likely member of the targeted population has been identified.

As implausible as this technology may sound, documents leaked to the *Washington Post* show that Chinese tech giant Huawei, working in conjunction with FRT startup Megvii, first developed and deployed a Uighur alarm system for Chinese police as early as 2018, and both companies later acknowledged the accuracy of this report.[11] In the years since, there has been abundant evidence that these companies and others have continued to develop ethnic alarm systems and other surveillance products using genetic and photographic biometric data, and that they have sold their products to repressive regimes around the world. The fact that there's a global market for such a technology is testament to the durability of the racialist and nationalist philosophies that led to some of the world's greatest tragedies over the past century. And the fact that technological development has so vastly outpaced social progress suggests that, unless we intervene proactively to regulate the development, sale, and deployment of such technologies, greater tragedies are likely yet to come.

BEWARE OF GEEKS BEARING GRIFTS

Not all uses of biometric technologies are as straightforwardly altruistic or dystopian as the cases we've described above. Sometimes, a seemingly beneficial application of biometric technology can obscure a much more sinister—or, at least, ill-conceived—purpose. Just like the wooden statue of a horse built by the wily ancient Greek general Odysseus and offered as a tribute to the Trojans, biometric applications presented in the guise of a gift or a convenience may contain far more dangerous possibilities within.

It's not always easy to spot the difference between a lovely statue of a horse and camouflaged troop transport. Furthermore, when a gift appears, unbidden, on your doorstep, it can be difficult to refuse. That's why it's generally a good idea, when presented with a new biometric technology application (or any new technology, really), to start with the question of what could possibly go wrong and then proceed as if that possibility were bound to happen.

Let's take the Moscow metro. In the fall of 2021, the Russian subway rolled out a new program called FacePay at all of its 241 stations and fourteen transit lines, the world's first large-scale system using FRT as a payment mechanism. Once a passenger registered their banking information and uploaded a selfie to the metro's mobile app, they could enter any station at any time, paying for their trip just by looking at a video camera mounted near the turnstiles. In the fanfare surrounding its launch, FacePay was touted as a modernizing convenience by the city's mayor and transportation department officials, who predicted that it would ease congestion by speeding up the flow of pedestrians through the transit system—a claim that was repeated uncritically in some news coverage.[12]

While this prediction may be true, it would also be naive to believe that making the subway system more efficient is the only way in which this new biometric surveillance system might be used. Russia is an authoritarian state that has deployed FRT on hundreds of thousands of public video cameras in its capital city alone, and it has a documented record of using that system to identify and intercept political targets, including opposition party members and human rights protesters.[13] Therefore, it seems more than likely that FacePay will also increase the efficiency with which the Russian government keeps tabs on its population, especially those it deems threats to its one-party political regime.

For the moment, Muscovites have at least a nominal choice not to register with the FacePay system, though this will likely change in years to come, and won't prevent the FacePay cameras from identifying them anyway. But citizens of the United States were not offered a similar choice when the Internal Revenue Service announced, a month after Moscow's FacePay launch, that it had unilaterally adopted a third-party commercial FRT system called ID.me and would require every taxpayer to register for the service with a selfie before being able to file their tax returns online. As with FacePay, government officials touted the new biometric surveillance policy as a modernizing move towards efficiency, writing in a press release that it would "improve identity verification" and "help taxpayers and the tax community"—another claim that was repeated in news coverage without adequate criticism or skepticism.[14]

Just as in the case of FacePay, many who study the intersection of technology and politics expressed deep concern about what could go wrong if the IRS moved forward with its plans to require an FRT-screened ID.me account for online taxpayers. Not only would it compel millions of Americans to add their faces to a privately owned FRT database, it would potentially introduce new legal and economic risks for taxpayers in the inevitable cases of the system not working as promised—for instance, potentially failing to accurately identify a taxpayer trying to file a last-minute return. After several months of voluble criticism from public-interest advocates, the IRS announced in 2022 that it would expand the range of options for online taxpayer identification, including—temporarily, at least—cutting FRT out of the equation altogether.

We are not suggesting that every use of biometrics in the name of the public good is undertaken with ulterior motives in mind. While it seems likely that the Moscow metro FRT system was built as much for the purposes of political repression as transportation efficiency, it's at least plausible that the IRS decision to choose ID.me was a genuine effort to prevent tax fraud and streamline the filing process. But that doesn't mean the risks to the people whose faces are biometrically captured are any lower; as we've argued throughout this book, once data are collected, stored, and processed, there's no telling where their secret lives will take them.

In fact, there are plenty of cases in which biometric data collected by one party for a certain purpose end up in the hands of a completely different party and put to other uses. To name one particularly egregious example, during its two-decade war in Afghanistan, the US military assembled a biometric database containing detailed information such as fingerprints, faceprints, and retinal scans for about twenty-five million Afghans— roughly four-fifths of the country's population. At the same time, the United Nations Refugee Agency also collected biometric data from Afghans, making aid contingent on participation in the database. While several privacy rights and human rights advocates voiced concerns about these practices, both the US military and the UN continued to insist that their biometric surveillance would aid in securing peace and security for the Afghan population.

Whether or not this was true at the time of the American mission, it certainly was no longer the case once the United States decided to pull up stakes and leave the country. As investigative journalists at the *Intercept* first reported a month after US troops withdrew in 2021, the Taliban had already seized the devices used by the US military to match Afghan civilians to records in the biometric database, potentially revealing the identities of countless Afghans who aided the American war effort and therefore exposing them to retaliation and interrogation. As Welton Chang, an Army veteran and chief technology officer of advocacy organization Human Rights First, told the *Intercept*, this was the result of a complete lack of foresight by US military leaders, who either didn't consider the possibility of data's secret life, or didn't care enough to take proactive measures: "I don't think anyone ever thought about data privacy or what to do in the event the system fell into the wrong hands."[15]

It is important to recognize that this absence of foresight by the US military was in no way unique or unusual. One of the key points we aim to make throughout this book is that, despite numerous voices clamoring for all of us to consider what might go wrong in the future, and numerous examples of what's gone wrong in the past, most institutions that build or use data collection and analysis technology are woefully limited in their ability to anticipate the long-term consequences of their actions. And this is not merely a failure of imagination. It's a failure of will, a combination of short-term utilitarian thinking and an endemic ethos of "it's better to ask forgiveness than permission." Therefore, we should not expect these outcomes to improve until someone with actual power—a national government or international standards body—creates a policy of accountability that requires technology builders and users to clean up their own messes.[16]

SECRETS AND REVELATIONS

As the rapid expansion of biometric data collection and analysis thrusts intimate details about our bodies into more and more private hands and in front of more and more public eyes, it is also helping to change the way we understand ourselves and our intimate relationships with others. Biometric surveillance is altering the topography of our lives, requiring us to see ourselves simultaneously as subjects and objects, and to constantly imagine ourselves as we are being seen through the sensors and algorithms of the ubiquitous data networks that surround us. We call this weird new state of mechanical double consciousness *algo-vision*, and we'll talk about it a lot more in the next chapter.

For the moment, we will focus on two specific ways in which biometrics have already radically altered our understanding of our public lives and private identities: deanonymization and the growth of the family tree.

We've already discussed the threat of deanonymization that emerges when several ostensibly anonymized datasets are combined and analyzed together. This has been an issue on civil liberties advocates' radars at least since the computer scientist Latanya Sweeney deanonymized Governor William Weld's medical records in the 1990s. But three decades later, the volume of biometric data collected and the power of analytical algorithms have both grown so precipitously that we can no longer safely assume that *any* information about our bodies, once captured by a sensor, will remain anonymous for long.

Facial recognition technology, of course, is by definition a form of deanonymization because it matches a photograph to an identity. But a variety of new tools are speeding up and enhancing this process. For years, companies like Apple and Facebook automatically ran FRT on every photo uploaded to their services, prompting users to verify the identity of people tagged by the ML algorithms (which, in turn, was used to make those algorithms more effective). In addition to public-facing features like this, companies such as Microsoft and Facebook also matched the faces in the photos uploaded to their servers against a database of missing children. They did so without telling their users, including this function in their terms of service, or advertising this social benefit in their marketing materials.[17]

The public FRT tools became so ubiquitous and accurate that they sparked a backlash, including class action lawsuits alleging that they violated biometric privacy laws. In 2019, Facebook attempted to stem the backlash by expanding the range of user options for FRT, including giving users the ability to opt out of automated tagging in publicly available photos—but not, apparently, the ability to opt out of FRT altogether (or, presumably, to opt out of back-end FRT functions such as matching with the missing children database). At the time of writing, it is reasonable to assume that any photograph that includes your face—whether it's a posed selfie or you're just a face in the crowd on the street or at an event—will be uploaded to an internet server and matched with your identity, with or without your knowledge and consent. It's also reasonable to assume that data brokers are becoming increasingly adept at using those algorithmic face matches to augment their profiles of you, including by logging the location-based metadata embedded in many of those photos.

Meanwhile, FRT developers say they are becoming increasingly successful at making matches based on scant data. When the COVID-19 pandemic emerged and billions of people around the world started wearing masks in public, facial recognition companies like Clearview AI and NEC responded by promoting new "mask removal" tools that promise to identify people who are covering their mouths and noses. This may be especially concerning to the countless millions of political protesters around the world who have worn such coverings to conceal their identities from the authorities.[18] And deanonymization of FRT has even gone one step further: computer scientists at the University of Caen Normandy in France have developed new tools that can reverse engineer the results of facial recognition software to identify the training data used for the software's underlying ML models.[19] In other words, they can look at the fake faces generated by a GAN like This Person Does Not Exist and reverse engineer the system to reveal the real faces that were used as seeds for the program.

Facial recognition is just one of many such examples of deanonymization. Other biometric data are also increasingly likely to be deanonymized and combined with unrelated databases, with problematic repercussions. This can include very sensitive personal data. For instance, in 2022, San Francisco district attorney Chesa Boudin publicly acknowledged that his police crime lab had been taking DNA collected from sexual assault victims in the course of police investigations and cross-referencing it with a separate database of criminal suspects, leading to at least one documented arrest of a rape victim for an unrelated property crime.[20] Though the case against this particular victim was dropped after privacy advocates questioned the ethics and legality of the practice, the damage was already done. People who suffer a sexual assault may now think twice before sharing their genetic information with police, lowering the odds that the people who attacked them will be brought to justice.

The widespread collection and analysis of genetic data is also changing how people think about their families. For millennia, the prevailing wisdom was best summed up by Telemachus in *The Odyssey* ("It is a wise child that knows his own father") and Launcelot in *The Merchant of Venice* ("It is a wise father that knows his own child").[21] Even in popular TV shows like *The Young and the Restless* and *Game of Thrones*, the mystery of key characters' parentage has been a common plot point. In other words, every family has its secrets, and paternity has always been more an article of faith than a statement of fact.

DNA analysis has turned this logic on its head. It has helped to unearth ancient secrets—for instance, establishing unequivocally that Thomas Jefferson fathered children by his wife's sister Sally Hemings, an African American woman whom he enslaved.[22] The Jefferson family and others had staunchly denied this claim for centuries. It has also exposed some modern secrets, such as revealing that dozens of fertility doctors in the United States have fraudulently used their own sperm to impregnate hundreds, if not thousands, of unsuspecting women.[23] While these particular stories were sensational enough to make headlines, there are countless other cases in which DNA databases have unearthed family secrets that didn't warrant news coverage, but still managed to turn individual families upside down. We interviewed someone who has experienced this personally: Maggie Clifford, a professional musician and climate communication scholar from Florida.

Clifford told us that several years ago, she built a free profile on Ancestry.com, a popular genealogy website, in order to research her family history. While she didn't submit her own DNA for testing, she added about ten generations' worth of lineage information that had been passed down to her by her mother. Several other family members did contribute DNA to the Ancestry.com database, however, including her brother and some of her father's siblings.

Not long afterwards, several of her family members received emails from a man we will call Tony, who had been raised in an adoptive family and was using Ancestry.com to find his biological relations. While Tony showed a genetic match with many people in the Clifford family, a quick process of elimination led to the conclusion that his biological father was Maggie's own father, making Tony her half-brother. As it turned out, Tony was conceived while Maggie's father—a strict Catholic—was still a freshman in college, and was born to his ex-girlfriend, months after she'd broken up with him, likely returned to her hometown in the Midwest, and cut off all communications. Nobody in the Clifford family had even known of Tony's existence until he reached out.

Maggie's feelings about her newly discovered brother and the process by which she learned of his existence have evolved over time. At first, she told us, her reaction was "Cool! How exciting!" But very quickly, she came to realize that her father did not share her unbridled enthusiasm. For him, this revelation altered his self-image, complicated his religious beliefs, and raised concerns about whether his existing family dynamics might be forever changed. These concerns tempered Maggie's own excitement and prevented her from reaching out to Tony independently of her father.

Two years later, Maggie entered a graduate program where she started studying subjects related to data privacy, and that led her to another way of thinking about the situation: neither she nor her father had submitted their DNA to Ancestry.com, but both of them had to deal with the emotional impact of Tony's appearance in their lives, due to their family members' unilateral decisions to share their own genetic data. "So their decisions—which, obviously, we didn't have to consent to—radically changed my understanding of our family and my dad's self-history," she said. "It felt like an assault on my father's privacy had been committed. An irrevocable one."

To Clifford, this event points to limitations in our understanding of the role of biometric data in our lives, in the policies that govern the uses of such data, and in our ability to make ethical decisions about those uses. While Ancestry.com markets itself as a way to find "distant past relatives," she said, the reality is that it's just as likely to reveal living family secrets, without seeking permission from everyone who is likely to be affected by such a revelation.

This fact, in turn, requires us to change our expectations about long-standing social institutions, like adoption. "Anonymous adoption policies can't exist any more in this context," she pointed out. Yet she also acknowledged that while this may be traumatic for people like her father, who learn late in life that they have children they've never met, it's also potentially empowering for adoptees that lack other information about their birth parents, who can now discover far more about themselves and their genetic histories than was previously thought possible.

After working through her conflicting feelings about this episode for several years, Clifford finally reached out to Tony. He never replied to her, but one of his adult daughters did, and today, Maggie and her newfound niece follow one another on social media. "I'm really grateful," she told us. "I feel like my life is richer from knowing that they are out there. Maybe one day, they'll need or want to have a connection with somebody who has a familial link with them—and I will be ready and willing to be that. So it is very complicated."

DEEPFAKES: BIOMETRIC PUPPETRY

As we've discussed, biometrics are frequently collected and analyzed in ways that may contribute to unpredictable social consequences. But, like other forms of data, biometrics are also *generative*, begetting new forms of data and information, which go on to have secret lives of their own in a never-ending, iterative process.

One of the ways in which biometric data have been used generatively in recent years is to fuel a new category of manipulated media that are known informally as *deepfakes*. You've probably read about them, seen some examples, and maybe even made some yourself. At the most basic level, a deepfake is a piece of audiovisual footage that purports to show an actual person saying or doing something they've never said or done, performing without consent according to the whims of its creator, like a digital puppet.

There have been several popular applications for creating deepfakes, many of them freely downloadable apps for smartphones and tablets. One of the first was the social media platform Snapchat's "face swap" feature, which became an internet sensation in 2016 by allowing users to take a photo or record a video while replacing their own face with that of a friend (or a stranger). Another popular app, Deep Nostalgia, released by genealogy platform MyHeritage in 2021, allows users to animate photographs, reconstructing faces in three dimensions based on AI analysis of two-dimensional imagery and adding lifelike gestures and motions.

At the same time, a new category of so-called voice cloning apps, such as VocaliD and Resemble AI, allows users to take short clips of recorded spoken audio and use them to reproduce any voice, in any language, using any textual input. When used in conjunction, these new animation and voice cloning apps have given billions of laypeople (not to mention professional media makers and manipulators) a production tool that even Hollywood studios lacked a decade earlier: the ability to generate a convincing televisual performance by any given individual without ever putting them in front of a camera or a microphone.

The term *deepfake* dates back to 2017, when an anonymous Reddit user with the handle Deepfakes first released a piece of hacked-together open-source computer code to the social media platform. The software allowed users to freely and easily map any photo of a face onto a video clip of another person's body. Deepfakes advertised its power by releasing several pornographic clips featuring the artificially superimposed faces of famous actors. Users of this software rapidly began posting deepfaked pornography and other types of video featuring both celebrities and private individuals appearing in compromising or embarrassing situations. This set off a firestorm of public concern and criticism, and platforms from Pornhub to Facebook vowed to eliminate deepfakes from their services—a pledge they lacked the technological expertise to support fully, even as dedicated deepfake-focused websites continued to proliferate beyond their digital borders.

Nonconsensual sexual imagery isn't the only potentially dangerous application for deepfakes. In the wake of global, politically volatile disinformation campaigns on social media in the mid-2010s, deepfakes were quickly understood to be another weapon in the media manipulation arsenal, with potentially deadly implications (imagine a fake video of a head of state declaring war on an ally). Journalistic outlets and democracy watchdogs rushed to warn the public against this threat, releasing obviously deepfaked videos of Donald Trump, Hillary Clinton, Boris Johnson, and other prominent political figures to demonstrate the power of this new tool. A cat-and-mouse game rapidly emerged within the programming world, with computer scientists at big tech firms and research universities developing new deepfake-spotting tools, and other programmers using software tools such as cycleGAN to make more sophisticated deepfakes that could evade automated detection.

In recent years, deepfakes have been used as potent propaganda tools, including by Russia and Belarus, which released doctored videos and imagery to build support for their unprovoked invasion of Ukraine in 2022.[24] But some technology and political analysts have pointed out that the rise of deepfakes poses a danger even greater than convincing people that false videos are true—namely, making people wonder whether true videos are false.

We interviewed Sam Gregory, the program director of Witness, a nonprofit organization that uses video and other media tools to document human rights abuses around the world. He told us that the public hand-wringing over political deepfakes has the unintended consequence of increasing the "liar's dividend," giving war criminals, authoritarians, and exploitative industries a plausible basis for denying the veracity of videos documenting their human rights abuses.[25]

Gregory gave us a concrete example of this concept in action: a 2021 video from Myanmar in which a former official imprisoned by the military government made some damning claims about corruption by Aung San Suu Kyi, the deposed head of state. Because the video was grainy and slightly out of sync, many supporters of the old regime claimed it was fake, and an automated deepfake detection algorithm seemed to support this assessment. But when Gregory teamed up with some forensic specialists to take a closer look at the video, they came to the opposite conclusion. While the corruption claims themselves may have been trumped up, and the "confession" may have been coerced, their analysis strongly suggested the footage was authentic. Though skepticism of any video released by the coup regime was certainly warranted, Gregory said, the rush to label such videos as deepfakes ultimately undermines the basis by which journalists and human rights advocates can hold such regimes accountable, because it treats documentary footage as presumptively artificial. Now, bad actors caught on camera can simply say, "No one can prove it's true, [therefore] it must be false," he told us.

The liar's dividend problem with deepfakes is clearly a serious social and political challenge, but it is merely one example of a much larger set of problems that will only increase as biometric data collection expands and the power of AI and ML algorithms continues to grow. The more that our bodies are scanned and analyzed, the more potential there is for generativity—for fake and manipulated digital spoofs of real, living people—and the more these spoofs will, in turn, call into question the legitimacy of authentic information about our bodies. Furthermore, as we rely increasingly on biometric data to understand ourselves and mediate our relationships with others, this threat becomes ever graver.

ALGORITHMIC BIAS GETS PERSONAL

In 2020, a Russian coder and self-professed "art-hooligan" named Denis Malimonov released a new piece of software that he called Face Depixelizer. Using a GAN, the software did exactly what its name implied: it took a low-resolution, pixelated image of a face and upscaled it to a "perceptually realistic" version of the image. Malimonov shared his code on social media, using an animated GIF to show the software at work, demonstrating how a blurry, pixelated image of a face was automatically reinterpreted by the software as a high-resolution image of a young woman with Asian features.[26]

The following day, a Swiss meteorologist named Cédric Sütterlin replied to Malimonov's post on Twitter, sharing a JPG image file that appeared to show a low-resolution (but easily recognizable) photo of US president Barack Obama—the first African American to hold that title—being resolved by the software into a high-res image of a white man with straight hair.[27] Sütterlin's only comment accompanying the photo was a short string of "thinky" emojis

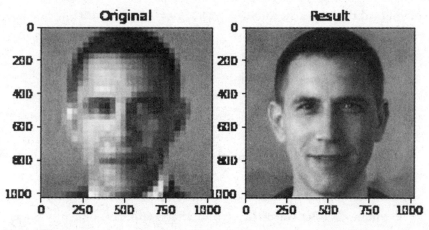

Figure 5.2
Screen shot depicting Obama portrait fed through Face Depixelizer. *Source:* Image by Cédric
Sütterlin.

(😅😅😅), but his tweet was "liked" over 22,000 times, suggesting that most people who saw it could clearly read the subtext: Face Depixelizer was one more in a long string of examples of AI suffering from algorithmic bias.

We discussed algorithmic bias earlier, demonstrating how software like GPT-2 "learns" to complete sentences with racist, sexist, and homophobic assertions due to the inherent biases in the datasets used to train it, as well as the biases of software engineers and end users themselves. Unfortunately, this problem is also rife in the world of biometric data and the algorithms that rely on them, which both compounds and complicates the many ways in which human bodies are already subjected to historically rooted systems of injustice. And, because computers and software are still frequently presented as objective, mathematically precise analytical tools, these biased biometric systems may also *normalize* racism, misogyny, and homophobia, making them seem natural or inevitable by obscuring their mechanisms of power in a maze of computer code. Few may have recognized Sütterlin's point, for instance, if Obama's face weren't so famous.

Face Depixelizer's whitening of the Obama portrait is the rule, not the exception, when it comes to FRT and GANs. Foundational biases baked into the training datasets (and into the field of computer vision as a whole) have manifested themselves in countless ways, from Google photo software that erroneously tagged Black people as "gorillas" to Twitter photo-cropping algorithms that privileged white faces over Black ones, and men's over women's.[28] But as we have discussed, FRT is just one of many biometric technologies that now play a ubiquitous role in our daily lives. Each of these other biometric systems, though their outcomes may not be as visible, shares a similar risk of amplifying and normalizing bias.

One of the less visible forms of biometric surveillance is speech recognition. As we described above, the power of this technology has grown significantly in recent years, to the point where voice-cloning tools are a basic commodity. Yet voice-activated interfaces still have difficulty understanding accents they aren't programmed to recognize, which can exacerbate social challenges faced by immigrants and other multilingual people, or even people with ethnic and regional accents that are underrepresented in training data and among programmers. Sinduja Rangarajan, a multilingual journalist whose first language is Tamil, wrote an investigative article in English for *Mother Jones* in 2021 that used her own frustration with Google Home devices as a jumping-off point to discuss the broader social challenges surrounding voice recognition, and critiquing her digital assistant through a postcolonial lens. As she framed it, the biased device had become "an intruder that was telling us how we should speak our own language in our own home."[29]

Like FRT, voice recognition can be a generative technology. Sometimes, the new applications built on its foundations don't merely reproduce social bias—they actually treat people's biometric traits as problems to be "fixed." For instance, in 2022, a California tech company called Sayso unveiled a new software application that it calls "accent transformation," which uses proprietary algorithms to alter people's voices so their accents will more closely match those of the people they're speaking to.[30] The company's website promotes the software as a way to improve speech comprehension to boost productivity in business communications.[31] While it may end up having this effect, that won't be the only consequence. The erasure of biological difference has a long and problematic social history. Especially for those of us who have been told that our bodies are somehow deficient because of their ethnic idiosyncrasies, it's difficult to view "accent transformation" separately from other "corrective" cosmetic interventions like hair straightening, skin lightening, and rhinoplasty—all of which treat white, Western European traits as neutral and normative, and other traits as unfortunate aberrances to be tamed.

The principal use of biometric data is still in the field of health and medicine, which means that the inherent biases baked into the collection and analyses of these data can mean life or death for patients whose doctors or insurance companies are relying on sensors and algorithms to make decisions about their health care. For instance, a 2021 study published in the *Journal of Internal Medicine* analyzed the records of over 56,000 patients with chronic kidney disease. Researchers found that "race multipliers" built into the algorithms evaluating patients for treatment had reclassified more than one-third of Black patients to a less severe status than their actual health warranted, diminishing their chances for lifesaving transplants, and thereby "exacerbat[ing]" racial "inequities" in the United States.[32] As with FRT and voice recognition, the inherent biases in these health care algorithms are the rule, not the exception. Another 2021 study, published in the *British Medical Journal*, showed what a ubiquitous and stark problem this is in the face of our current global pandemic. The study's authors reviewed 232 diagnostic and prognostic models related to COVID-19 and found that 97 percent of them presented a "high risk of bias," with the remaining 3 percent "unclear."[33]

Whether depixelizing a selfie, using voice commands to play "Despacito" on a smart speaker, or evaluating candidates for kidney transplants, the biometric data systems that now surround us have reproduced historical biases and inequities—and threaten to make them worse unless we take aggressive steps to change our technologies *and* our societies. Yet, at least these examples offer some concrete social benefits, as well, such as making consumer technology more accessible for disabled users, and helping health care providers evaluate the severity of diseases. Unfortunately, however, some uses of biometric data just skip the good stuff altogether and go straight to the bias. We think of these biometric applications as a kind of digital snake oil.

Digital snake oil isn't always easy to spot. There are so many mind-blowingly exciting new breakthroughs in the world of data and technology happening on a near-daily basis that even jaded tech journalists might be forgiven for believing and reproducing some nonsensical hype once in a while. But it turns out to happen a lot; thankfully, there is an entire cottage industry of technologists and tech ethicists on social media platforms like Mastodon, YouTube, and TikTok who spend their days debunking some of these ridiculous and potentially harmful claims. Nonetheless, most of the snake oil ends up making it into public consciousness (and sometimes, the marketplace) without anyone raising a red flag along the way and saying, "Hey, maybe don't do this."

We've already mentioned some examples of digital snake oil in this chapter, such as the DNA phenotypes developed by the Chinese government to create their potentially genocidal "Uighur alarm." Nearly every scientist who understands how genetics actually work will tell you that extrapolating what a person's face looks like based on their genetic profile is virtually impossible. Even if you get a rough approximation, it will produce thousands of false positives. That's because, as with so many other algorithmic processes we've described in this book, "ethnic alarms" are fundamentally probabilistic models, so their output can't be reasonably interpreted in terms of a definite, binary yes or no. To put it another way, the best such algorithms can offer is the prediction that a given person has some likelihood of looking a certain way based on their DNA. Sadly, this limitation hasn't stopped a raft of companies from developing products that make absolute, binary claims about accuracy, or from hawking these products to law enforcement and other government agencies in the United States and elsewhere.

Other equally absurd and potentially dangerous examples abound. For instance, in 2017, a computer scientist from the University of Georgia teamed up with a business school professor at Stanford to write a paper (that has since been debunked) claiming that they had developed a neural network capable of identifying homosexual men and women based on a single photograph, with the accuracy increasing dramatically (to 91 percent for men and 83 percent for women) when given five images of the same person.[34] Leaving aside for a moment the demonstrable biases and limitations inherent in the data used to train the model, as well as the fact that sexuality encompasses a multifaceted set of identities, practices, and subcultural styles that can't be reduced in any meaningful way to simple gay-straight binaries, let's consider the problematic social implications of automated "gaydar" like this.

There are numerous societies around the globe, from Nigeria to Iran to North Dakota, in which millions of people have been persecuted, imprisoned, tortured, and even executed for their alleged homosexuality. What might the leaders of these societies do if empowered with a digital tool that made authoritative claims about the sexual orientation of anyone living within their borders? Even if these leaders were savvy enough to spot snake oil when they saw it, such software would provide a veneer of objectivity and authenticity to anti-LGBTQ bias, both normalizing the idea that gay and straight people are fundamentally different in some biological respect and giving authorities justification to persecute anyone singled out by the gaydar algorithm. Even if the software were used for ostensibly LGBTQ-friendly purposes—improving dating apps, or outing public figures who espouse homophobic policies—it would still serve to validate the legitimacy of such software, inevitably leading to more persecution in the long run.

It gets worse. A whole new category of biometric technologies purport to measure temporary, subjective aspects of a person's body and identity, rather than more durable or objective ones. One such variety of snake oil is biometric lie detection. Unlike traditional polygraph tests, which measure blood pressure, breathing patterns, and the amount of sweat on the skin to infer the truthfulness of people's statements, a new species of lie detection software relies on digital biometric tools like eye tracking and FRT to observe minuscule changes in people's expressions, which are then correlated with the accuracy of their claims. Although thousands of law enforcement agencies and other customers around the world have adopted these new biometric systems, there is good reason to be skeptical of lie detection technology, both old and new. As the American Psychological Association has written, we don't even have a clear understanding of how deception works in our bodies and our minds, and therefore "without a better theoretical understanding . . . lie detection technology seems highly problematic."[35]

There are many other similar kinds of digital snake oil, all proposing to use biometrics to track the passing tides of our thoughts and feelings. Many fall within the broader category of affective computing, a legitimate and diverse field of research. An entire subfield, generally referred to as *emotion AI*, promises to measure people's true emotions (whatever that means). Others purport to measure adversarial intent—that is, to predict whether someone plans to do something bad. Some even claim to capture the images from people's dreams (this has been something of a holy grail in biometric research for decades, for some reason).

What do these systems have in common? First, with the exception of dream surveillance, they have all been adopted by powerful institutions to make important decisions about people's lives. Second, because these systems are probabilistic, the best they actually can do is offer a compelling guess at the truth, no matter how definitive their presentation seems to be. And third, unlike other biometric systems, they can't be refuted definitively, so there's no recourse for the people whose lives are altered when data about their bodies are captured, analyzed, and acted upon. If an algorithm says you're feeling happy or sad, or that you intend to do harm, there's not much you can do to rebut it other than to say, "No, I'm not." And that assumes the people using the algorithm in question bother to tell you what they've done before arresting you, rejecting you for a job, or denying you a lifesaving health procedure.

Although DNA phenotypes, algorithmic gaydar, emotion AI, and other digital snake oils are all on the cutting edge of balderdash, thanks to their reliance on rapidly developing technology like digital sensor networks and machine learning algorithms, it is important to understand them in a historical context as the latest in a long lineage of pseudoscientific tools whose principal function is to legitimize bias, exploitation, and subjugation by giving them a sheen of objectivity and empirical weight. As Sharrona Pearl, a historian of biometrics and professor of medical ethics at Drexel University, explained to us, the drive to develop technology that would reveal people's personalities, emotions, and intentions dates back to the nineteenth century.

At that time, the rise of modern cities meant that many people in urbanizing societies found themselves surrounded by anonymous strangers instead of familiar faces. Originally, she told us, a concept called *physiognomy*, which held that people's physical features revealed their moral natures, served as a kind of "shortcut to individualization" for people living in these cities. We might not know one another as we pass on a crowded street, she explained, but thanks to physiognomic principles, "I could look at your nose and decide whether or not you're trustworthy."

From the very beginning, physiognomy was inextricably linked with other dubious scientific practices and premises, including those that purported to demonstrate the supposed racial superiority of one group over another, and those that claimed women were less intelligent than men. It also led many people to alter their appearances in order to conform to supremacist standards of beauty and physiognomic principles of morality, and in so doing gain social status. "So it's very tied into class mobility," Pearl said. "You know, the 'pulling yourself up by your bootstraps' self-improvement narrative."

Pearl's research demonstrates the direct progression from nineteenth-century physiognomy to twenty-first-century biometrics, both driven by the same kinds of pressures and anxieties. If the rise of the modern metropolis put everyday people in contact with strangers for the first time, the creation of the global internet has amplified that process a thousandfold. For billions of people today, it's virtually impossible to attend school, have a job, read the news, find a date, see a doctor, or even watch a movie without using online apps and tools—and that means opening parts of our lives to the billions of other people doing the same. So it's only natural, Pearl concludes, that we'd resort to the same kinds of techniques to help us manage our insecurities, relying on algorithms to help us sort all these strangers into categories such as "friend" and "foe," "like" and "unlike." Like their predecessors, what these new forms of snake oil lack in rigor is made up for in hype. Each one is billed as "revolutionary," an "innovation" that will somehow fix the world's problems, even if the problems' causes have nothing to do with technology.

Unfortunately, these new biometric systems are just as flawed as their predecessors, and we end up replicating our biases and hiding them from ourselves beneath a veneer of scientific objectivity. "All these empiricist notions of the body are attempts to get out of structural limitations and biases," according to Pearl. "But of course, not only do they reproduce them, but they actually are a function of them."

These problems are likely to compound as biometric technology continues to be developed and deployed, propelled by billions of dollars in annual investment. Though there's some pushback (such as the efforts to mitigate and regulate facial recognition technology), it's dwarfed by the overall momentum of biometrics. Emerging fields, from nanobiology to neurotechnology to the metaverse, will continue to open up new horizons for science and pseudoscience alike, while opening our bodies to even greater biometric surveillance. And these developments won't simply change the way we understand and interact with one another—they are already fundamentally altering the way we understand ourselves.

6 THE OVEREXAMINED LIFE

On a clear day, you can see a lot from the corner of Broadway and Van Ness in San Francisco's Polk Gulch neighborhood. The gothic spire of St. Brigid Church looms over the intersection, and Van Ness itself slopes down gracefully toward the bay, where Alcatraz Island sits just off shore. Though Alcatraz has been a national park for decades, its dark history as a military prison and federal penitentiary still looms over the island—a reminder of how important and fragile freedom can be, and how every generation must work to further the causes of liberty and justice.

Looking out from high on the hill above the bay, you may find the poetic phrase "standing on the shoulders of giants" coming into your head. But what you might not realize is that you're also literally standing on the shoulder of a giant portrait of Stacey Abrams, the modern-day civil rights icon who is a mainstay of Georgia state politics and a national advocate for free and fair elections. There's nothing on the street to mark the spot—no plaque, no paint, no commemorative plate embedded in the sidewalk. To see Abrams's portrait, you'd have to open Strava, a popular fitness-tracking smartphone app, and look at the San Francisco city map. In 2021, a cyclist named Frank Chan chose to express his admiration for Abrams by plotting a fifty-kilometer, six-hour bike route around the city that, when rendered as a continuous line on the map, would appear as a drawing of her that spanned six and a half miles end to end.[1]

There's nothing new about people making giant drawings using the landscape as a canvas. Two millennia ago, the Nazca people in southern

Figure 6.1
A Nazca spider, the Cerne Abbas Giant, and Frank Chan's Abrams portrait. *Sources:* Diego Delso, Dorset Council, Strava.com.

Peru drew dozens of such geoglyphs, representing animals and plants like spiders, cacti, and monkeys with lines sketched in the desert a quarter of a mile long. And around one millennium ago, give or take, some Saxons used chalk rubble to draw a 180-foot-long club-wielding figure on the side of a hill in Dorset, in the south of England. Long before the invention of powered flight, human beings were already capable of imagining what the world must look like from above and creating art that could only be seen in its entirety from such a vantage point—even if nobody had the means to reach a sufficient altitude to see it. But what makes Chan's drawing different from its ancient antecedents is that his portrait requires a double abstraction: picturing the world from high above and imagining it through the eyes of an algorithm.

We've noted the pervasiveness of data in our lives, including the digital devices, information networks, data industries, and invisible algorithms that surround virtually every space, place, and action in contemporary industrialized society, from our public and professional spheres to our private and intimate ones. But while we've talked a lot about how this new digital surveillance infrastructure alters our social lives and affects how others see us, in this chapter, we'll shift perspective and focus on how ubiquitous data alters our inner lives and affects the ways we see ourselves. Most of us don't bike fifty kilometers up and down the hills of our hometowns to pay homage to our favorite political figures, but each of us, in our own way, is a lot like Frank

Chan: we've learned to see our worlds through the eyes of the algorithms that surround us, and we've altered our habits, our behaviors, our identities, and sometimes even our bodies in response to what we see reflected and refracted through those data interfaces—and what we believe others might see when they look through them at us.

QUANTIFYING THE SELF

Humans have probably been measuring themselves for as long as they've been measuring the world around them. Even those of us who grew up without smartphones or the internet took it for granted that stepping on a scale each morning would help us keep track of our weight, sticking a thermometer under our tongue was the best way to diagnose a fever, and standing against a doorjamb to mark our height each year was an essential rite of passage as we progressed through childhood. Sometimes, we even repurposed tools that weren't intended for self-measurement; one of the authors of this book had a great-grandfather who worked as a butcher and weighed himself regularly, well into his nineties, by hanging from a meat hook.

We seem to have an inborn urge to collect data about our own bodies, using whatever tools and paradigms we have at our disposal. That urge is spurred on by social pressures and forces: the more we *can* know about ourselves, the more we *should* learn, we're told, for our own good and for the benefit of those around us. As medical ethicist Sharrona Pearl explained to us, these pressures date back at least to the 1800s and the rise of pseudosciences like physiognomy, which held that the measurable features of a person's body could be seen as a guide to their moral character. According to the backward logic of physiognomy, Pearl told us, people believed that changing the appearance of the body could also change a person's moral stature. This meant it was "actually your obligation" to look your best in public, using any means necessary, from makeup to surgical intervention. In her own scholarship, Pearl has shown how these ideas filtered down over the centuries into modern media and tech cultures, and how they still inform the narratives of popular makeover shows and cosmetic surgery trends in the age of social media.[2]

Self-measurement has changed a lot since the nineteenth century, and even in the two or three decades since the internet began to transform our social infrastructure. Standing on a scale or popping a thermometer in your mouth are immediate, local behaviors, discrete events requiring the volition of the individual, which offer data only to people in the direct vicinity. They're self-contained uses of biometric technology. But the growth of cyber-physical networks, the mass adoption of smartphones, and the widespread deployment of wireless internet each contribute to a new set of conditions in which continuous, networked measurement has become the norm. As we discussed in chapter 5, biometric devices and algorithms have rapidly proliferated in our public spaces, our homes, and our bodies, driven by commercial and state interests who seek to exploit the resulting data. Perhaps it was predictable that people surrounded by biometric networks would begin to use them for the purposes of self-measurement—and self-improvement—as well.

In 2008, two editors at *Wired* magazine, Gary Wolf and Kevin Kelly, started a blog called the *Quantified Self*, which aimed to track the myriad ways people were using cutting-edge biometric technology to learn new things about their lives and bodies.[3] As Wolf explained in a widely viewed 2010 TED talk, adherents of the burgeoning quantified self (QS) movement saw these new biometric self-tracking practices as a means to "self-improvement, self-discovery, self-awareness, [and] self-knowledge."[4] In other words, the QS movement, and the ubiquitous tracking technologies it relied on, were presented as unalloyed benefits for humanity and an important new path toward self-empowerment. In the years since then, public interest in QS as a movement has waxed and waned. Google Trends, a tool that tracks the popularity of search terms over time on the world's most popular search engine, shows searches for "quantified self" climbing precipitously for the three years after Wolf's TED talk, then tapering off slowly through 2018 or so, to a steady level at about one-tenth of peak popularity.[5]

There are probably several reasons for this. First, like all new ideas, QS had its moment in the sun when millions of people were first exposed to it and searched for more information; over time, it became just one more well-known concept in the lexicon of internet culture. Second, over the past decade or so, there has been a widespread "techlash" ("tech" + "backlash")—a popularized critique of the pervasive role that big tech and big data play in public life, and an increasing degree of popular mistrust of these technologies and the organizations deploying them. Finally, perhaps paradoxically, QS has waned in the public imagination precisely because the act of self-tracking using cyberphysical systems has become so normal that it's almost invisible to many of us.

According to market research firm IDC, about a billion wearable devices were sold in 2020 and 2021, and the market continues to grow in size by about one-fifth with each passing year.[6] According to analysis by the *Economist*, smartwatches had negligible adoption levels in 2015, but by 2022 well over 40 percent of US households had at least one—a growth curve just as steep as the adoption of the internet two decades earlier.[7] Chances are good that if you're reading this book, you wear a smartwatch, or use a fitness tracker, sleep tracker, medical implant, or other networked personal biometric device on a regular basis. If so, you may be contributing to the growth of QS without even knowing it.

QS is also evolving, reflecting broader changes in the way that biometric technology is designed and employed. While Wolf's 2010 TED talk focused on empirical, nuts-and-bolts measures of physical health and activity, such as sleep trackers and exercise logs, newer QS applications build on some of the digital snake oil technologies we discussed in the last chapter, fusing them with popular self-help and self-improvement trends to offer some fairly dubious services.

For instance, a digital wristband called Moodbeam features two buttons on its outside panel—a yellow one, which users are supposed to press if they're feeling happy, and a blue one, which users are supposed to press if they're feeling sad. The target customers for Moodbeam aren't the individuals whose emotions are supposedly being measured, but rather their employers, who are given a suite of tools to evaluate the overall happiness of their workforces. According to the Moodbeam website, over one hundred companies have used their tools to track the real-time emotional state of their employees.[8] It's difficult to say which is the most problematic aspect of this story: the reduction of the complex human emotional spectrum to a simplistic happy-sad binary; the coercive implications of an employer spying on the real-time emotional state of its labor force; the limited likelihood of workers accurately self-reporting under these coercive circumstances; or the fact that credible journalistic outlets like the BBC and the *Financial Times* covered the technology enthusiastically without raising any of these concerns in their articles.[9]

Another emotion AI company called Feel Therapeutics also claims to track people's emotional states using a biometric wristband. In this case, both the wearers and their employers are the intended customers. But nobody's going to pay for the privilege of pressing a "happy" or "sad" button to tell themselves how they feel (at least, we hope not many people would). So this wristband uses more ostensibly scientific measurements, such as pulse rate and skin moisture, to gauge changes in the wearer's mood. Then it asks users to choose an emotion that best matches their mood from a preselected menu, which Feel uses to build a psychological profile of the wearer. Finally, the company has a service component, offering a suite of possible behavioral interventions, from breathing exercises to a weekly fifteen-minute video chat with a licensed mental health professional. The company's website claims that the device will improve mental health care by replacing "subjective" self-evaluations by patients with "objective data," with the aim to "reinvent the way we diagnose, manage and care for Mental Health."[10]

There are countless other examples of biometric devices and services currently offering some kind of QS tracking-plus-intervention to individuals, employers, health care providers, and other interested parties—far too many for us to list, let alone discuss thoroughly, in these pages. But our point isn't simply that many of these individual devices are at best ineffective and at worst a danger to health care outcomes and civil liberties, though this is certainly the case. Our larger concern is that the aggregate effect of these biometric devices and QS services in our personal lives is a kind of datafication of lived experience. In the field of linguistics, an idea called the Sapir-Whorf hypothesis holds that the words we use don't merely describe our shared reality but also help to shape it. According to this theory, the structure of a language directly influences and constrains how its speakers experience the world around them. For example, people are much more likely to distinguish visually between a green object and a blue one if they speak a language (like English) that has separate words for *green* and *blue*.[11] Similarly, the more we come to understand our bodies, and even our emotional states, through the categories and quantities promoted by our biometric devices and the algorithms that accompany them, the more likely we are to experience our own lives according to those schema.

Not only might this reduce the range and subtlety of our sensory and emotional palettes to the minimum required by a given technology's business model (think of Moodbeam's absurdly simplistic happy vs. sad dichotomy, or Facebook's slightly more nuanced list of seven reaction emoji); it also introduces the possibility of false conclusions and self-fulfilling feedback loops. If you wake up in the morning feeling refreshed, then check your sleep tracker and discover it only gave you a 62 percent score for the previous night, are you more likely to doubt the (empirical) veracity of the device, or to reevaluate your own (subjective) experience? *Maybe I didn't sleep so well after all*, you may be tempted to tell yourself; *Guess I'll make a second cup of coffee.* Believe it or not, there's a name for this phenomenon: the *nocebo effect.* Just like a placebo sugar pill can trick people into feeling better if they believe it contains therapeutic drugs, a nocebo can do the opposite, making healthy people feel worse through the power of suggestion. Though the term was coined in the 1960s, long before smartwatches, recent research has demonstrated that the power of negative suggestion can apply to apps, much as it can to pills.[12]

Although this may seem like a trivial concern—maybe you'll feel a little more jittery than you would have without that unnecessary second cup of coffee—it can really add up, especially when we consider the countless devices and algorithms influencing the decisions of billions of people on a daily basis. On a global scale, even a slight tweak in individuals' perceptions and behaviors can add up to tectonic changes in our markets, our politics, and our cultures. And, increasingly, the companies that make and sell these devices and algorithms are aware of such consequences and willing to use them for their own gain.

DYSMORPHIA BY DESIGN?

"WE MAKE BODY IMAGE ISSUES WORSE FOR 1 IN 3 TEEN GIRLS"

Thankfully, this alarming claim isn't the marketing tagline for some villainous cabal of evildoers bent on destroying the emotional well-being of young women. But it is a very real self-assessment, by a very powerful organization, and its implications couldn't be more problematic if it were the outcome of a fiendish plot.

The headline in question is from an internal slide deck circulated at Instagram, a leading photo- and video-based social network owned by Meta, which also owns Facebook and WhatsApp. The presentation, circulated in November 2019, summarized internal research at the company, which assessed a broad range of psychological consequences for users of the service. In addition to the damaging body image issues, large portions of users surveyed said that other negative feelings had "started on Instagram," including "have to create the perfect image," "not attractive," "don't have enough money," "don't have enough friends," and "not good enough."

These findings echoed independent assessments of visual social media's role in the emotional lives of users that had been published by academic and commercial researchers for well over a decade. For instance, Renee Engeln, a psychology professor at Northwestern University, published an experimental study in 2020 showing that young women who had spent seven minutes using Instagram "thought about their appearance more, compared their appearance to others more, felt worse about how they looked, and reported more negative moods" than those who had spent seven minutes looking at Facebook instead.[13]

But while scholars like Engeln took pains to publicize their findings and to use them to fuel a public conversation about the role of social media in young people's lives, Meta kept the negative implications of its research findings private. Meta CEO Mark Zuckerberg even testified that social media had "positive mental health benefits" before the US Congress in March 2021.[14] The only reason the general public learned about Instagram's self-reported deleterious effects was because a corporate whistleblower named

Frances Haugen shared the slide deck in question with the *Wall Street Journal*, which published excerpts from it six months after Zuckerberg's testimony.[15] A few weeks later, Meta provided its own copies of the research directly to Congress and published the findings on its blog, along with explanatory text that aimed to downplay its troubling implications.[16]

Instagram may have dominated the headlines in 2021, thanks to Haugen's revelations, but it's far from the only internet platform that appears to play a role in shaping people's body image, and their self-perception more broadly. It's difficult to find a social platform that *doesn't* offer users the opportunity to change their appearance with the tap of a button, from the massively popular beauty-enhancing "face distortion" features on social media apps like Snapchat and Facebook that can sharpen your jawline, increase the size and roundness of your eyes, or plump up your lips in a matter of milliseconds, to more subtle features like the "touch up my appearance" setting on videoconferencing apps like Zoom and Google Meet, which can smooth out wrinkles and reduce the appearance of blemishes while leaving your basic facial architecture intact.

These "corrective" filters, which typically enforce a standard of beauty that is determined by designers and programmers in ways that reflect their own biases about race, gender, ability, and other historical vectors of prejudice (and which may be at odds with the aesthetics and values of users themselves), are not only easily accessible on such platforms, but are increasingly switched on by default. This is reportedly the case for many of the most popular social media platforms in China, including Douyin and WeChat. It was also temporarily true for the American version of Douyin (known as TikTok) for a while in 2021, though the company claimed it was an accidental glitch, and turned the "beauty" filter back off by default a few days after users first reported the issue.[17] Despite this slight misstep, most social media and online communications companies continue to promote these filters as vital elements of their platforms, and they frequently trot out data showing their popularity as evidence that they are merely responding to consumer demand.

It would be simplistic to conclude that social media just "make people unhappy," or that visual filters make users "hate the way they really look."

People suffered from body dysmorphic disorder (a compulsive focus on perceived flaws in one's physical appearance) and low self-esteem long before Snapchat, TikTok, and Instagram were household names. And there are doubtless plenty of users who find social media filters entertaining, or even affirming. The ways that people use new technology are always influenced by existing social forces and pressures.

Yet it would also be dangerously naive to assume that apps and filters are simply a new twist on an old trend, and that the root causes of people's anxieties and neuroses are completely unrelated to the distorted images they see of themselves and others via social media apps. Especially for young people, who are justifiably anxious about their physical appearance and its implications for their future happiness, the central role of social media in their lives and the widespread use of beauty filters by their peers contributes to a classic no-win situation in which they are forced to choose between participating in a digitally augmented aesthetic arms race, or risk losing social status.

As online video communication platforms become more deeply ingrained in our civic, professional, and cultural lives, it seems likely that the kinds of challenges Instagram identified for teen girls on its service will come to affect a broader portion of the overall population, with more far-reaching consequences. For instance, in 2020, when the COVID-19 pandemic first sent millions of office workers home for an extended period of time, and video-based meeting platforms like Zoom saw dramatic surges in popularity, one of the unforeseen consequences was skyrocketing demand for cosmetic plastic surgery. These two trends were clearly related. As the medical director of one surgical center told the *Washington Post*, during consultations with potential patients, "9 out of 10 people commented about noticing these things [they hoped to correct with surgery] over Zoom."[18]

THROUGH THE LOOKING GLASS, DARKLY

While the lines between webcams, beauty filters, social media, negative body image, and surgical intervention may be simple to trace, there are other, subtler ways in which the ubiquitous presence of digital sensors and algorithms are helping to warp our self-perception and identities. That's because most of the data collected about our bodies and behaviors aren't visual in nature, and most of the time, we don't get to see the computer's assessment of us rendered in real time on a screen.

Instead, our digital social environment is more like a series of one-way mirrors and weird echo chambers in which we can sometimes tell that we're being observed and judged, but are rarely able to audit (let alone correct, delete, or amend) the information associated with us via obscure algorithms. This limited visibility, in turn, can lead us to spend a lot of time and energy trying to figure out how to express or present ourselves in ways that will show us in the best light in this strange new environment. In effect, we are compelled to see ourselves simultaneously as active participants in our own lives, and as something akin to video game avatars, complete with stats, powerups, and cheat codes.

We don't necessarily do this on our own; entire online cultures emerge out of discussions about how to identify—and game—these informatic lenses, and how to turn their distortions into an advantage, instead of a liability. There are sections of the social media website Reddit dedicated to sharing hacks that will fool an exercise tracker into recording nonexistent activity (and thereby fool some insurance companies into offering lower rates or other incentives), such as attaching a smartwatch to a drill bit and spinning it rapidly. Even if we succeed in these efforts, however, we've already lost another battle. The moment we change our behaviors and appearances to serve the dictates of an algorithm, we subvert our own needs and desires to suit the values of whoever has designed or programmed it. This kind of behavior wouldn't and couldn't exist without the fractured state of consciousness that comes from our collective efforts to internalize the obscure logic of the data systems that pervade our lives. In this book, we refer to this information-lensed perspective as *algo-vision*.[19]

Algo-vision has become so omnipresent in our lives that it's difficult to choose just one emblematic example. Our spoken and written languages, while always in flux, have changed a lot in recent years, integrating bits and pieces of internet jargon (such as saying "LOL" instead of actually laughing), and also shifting in subtle ways that reflect the real and imagined roadblocks and potholes we face in our digital travels. As tech reporter Taylor Lorenz described in a 2022 *Washington Post* article, content filters and moderation policies on popular social media platforms have driven users to substitute euphemisms for banned or discouraged words—such as "unalive" instead of "dead," or "spicy eggplant" instead of "vibrator"—and those substitutions have taken on a linguistic life of their own, which Lorenz calls "algospeak."[20]

There's nothing particularly new about the observation that computer culture and online communications can alter the way we write, speak, and develop our identities. As MIT professor Sherry Turkle observed more than two decades ago, "One of the most important cultural effects of the computer presence is that the machines are entering into our thinking about ourselves."[21] Other turn-of-the-millennium researchers explored phenomena such as textspeak (like "IMHO" as shorthand for "in my humble opinion") and leetspeak (codes used principally by gamers and hackers in the '90s and '00s) and traced their evolution from chat windows into common parlance.[22] A decade later, social media researchers danah boyd and Alice Marwick reported that teens online were using "social steganography," a loose system of code words and oblique cultural references, to have private conversations in public.[23]

But what's new about algospeak is that, instead of trying to exclude outsiders (like leetspeak) or fool grownups (like social steganography), this time around, the changes in our language are the direct result of people trying to evade and game *algorithms*, based on a dim, collective sense of the arbitrary and ever-changing rules encoded in commercial communication platforms. This, in turn, creates a feedback loop: the more we become aware that our speech is being monitored and potentially censored, the harder we work collectively to bypass those new forms of control. In fact, algo-vision writ large can be understood as the result of such a loop. The more we become aware of networked algorithms in our life, the more we alter our collective behaviors, which feeds back into how the algorithms function, exaggerating and warping our shared social fabric so that our biases, idiosyncrasies, and dysfunctions stand out like the screech of a microphone held too close to a loudspeaker.

Another example of algo-vision is the way we now represent ourselves on the job market. This has always been a performative experience; whether applying for a high-ranking executive position or an entry-level job doing manual labor, applicants typically strive to put on their "best face," presenting an idealized version of themselves to the people in charge of hiring. But today, online recruiting websites and résumé-sorting algorithms have become integral to the hiring practices of thousands of companies. This means that, increasingly, job seekers must impress software before they have the opportunity to impress another human being.

Because we can't know what an algorithm "sees" as it scans our work histories and résumés, job seekers today are forced to rely increasingly on intuition and internet folk wisdom to make a more or less educated guess about how to stand out. This, in turn, changes how companies search for new workers, generating a feedback loop that distorts the hiring process in unpredictable ways. As long ago as 2008, linguistic scholars Nicole Amare and Alan Manning warned that the "formatting tricks" and "deceptive key-wording techniques" adopted by online job seekers presented both ethical and practical problems for the labor market.[24]

Yet these warnings and recommendations have been largely ignored, and by the 2020s, gaming résumé-sifting software became its own cottage industry, with companies like Jobscan charging millions of users $50 per month apiece to optimize job application materials to appeal to the algorithmic gaze. The end result is that applicants are less honest in their self-presentation, employers and workers are both less likely to find a good match, and the added costs job seekers must pay to game the system exacerbates the economic gap between haves and have-nots. People who already have some money are better able to get jobs, while the people who need money the most can barely get interviewed. It's not the fault of any single party, just the emergent consequences of introducing these sorts of algorithms into the hiring process.

Algo-vision doesn't merely tweak our languages and warp our markets; it's a fundamentally new paradigm that alters daily life and lived experience for untold millions or even billions of people, and we are only just beginning to understand its broader implications for human society. As in the case of creative bicyclist Frank Chan, it can be culturally generative, giving us a new perspective from which to engage in creative expression and participate in political discourse. It can also rewire our social relations and hierarchies; think of the rise of social media influencers, and the prevalent ethic of "doing it for the likes." According to this principle, any cultural behavior, from self-destructive activities like eating a spoonful of cinnamon to altruistic ones like attending a civil rights protest, can be incentivized by the prospect of garnering quantifiable attention and acclaim, which can be converted into quantifiable social and economic capital.

Algo-vision even influences the way we understand our bodies—not only intellectually, as with the QS movement, but somatically, in the sense of our physical experience of being alive. We interviewed Laura Forlano, a design researcher and social scientist who is disabled and wears a "smart" insulin pump to manage her type 1 diabetes.[25] She described for us the unnerving experience of relying on a networked biomedical device to maintain her health. In her words, "This AI system is keeping me alive, but it's also ruining my life."

Forlano calls her lifesaving implant "dehumanizing and intrusive," which is both her personal assessment as a user, and her professional assessment as a design researcher. "This particular system has so many alerts and alarms," she told us, that it woke her up several times per night, several nights a week, for years on end. None of these alerts and alarms were directly related to immediate threats to her physical wellbeing, and she believes she would have been healthier if she'd been given the option to turn them off. Unfortunately, the makers of the device didn't think to include that option for users, so she and countless other diabetics suffer from sleep deprivation because of the designers' shortsightedness (exacerbated, she says, by well-meaning regulations that shift the burden of responsibility for managing these devices onto users).

Even worse, Forlano told us, the smart device's frequent beeping and vibrating led her to start ignoring its alerts, which meant that she was less likely to know if something important was actually happening, such as a life-threatening drop in blood sugar levels. Forlano sees this phenomenon, which is often referred to as *alert fatigue*, across a broad range of biometric and other technologies. In the case of medical implants, her principal concern is that it "affects both doctors and patients," meaning that neither diabetics nor their caregivers can effectively discriminate between life-threatening and irrelevant data, so they end up ignoring both, with potentially tragic consequences.

Alert fatigue is a great example of what can go wrong with algo-vision. At first glance, the lifesaving capacities of smart insulin pumps and other biomedical implants make them seem like miraculous technologies from a *Star Trek*–style future. But when people begin to internalize the data readouts from these devices to augment and supplement the experiences of their own bodies, poorly calibrated interfaces and lazy design can make users panic unnecessarily, lose sleep, or even ignore grave medical threats as emergency alerts become indistinguishable from background noise. In essence, this is true in all cases of algo-vision, from the mundane to the catastrophic. When we rely on secret algorithms and proprietary devices created by self-interested third parties to mediate our health, identities, relationships, and subjectivities, we give up a tremendous amount of individual and communal power. This puts us at the mercy of the biases, beliefs, and blind spots of the designers and organizations that engineered the technology to begin with.

When we choose to use a given technology, Forlano told us, we tend to understand the choice as an aspirational declaration of "who we want to become"—just like choosing a new pair of shoes, or styling our hair a certain way. Marketing messages for technology tend to reinforce this perception (think of the famously effective "I'm a Mac, I'm a PC" ad campaign from the first decade of the twenty-first century). Yet, Forlano warns, technology use is complicated by the outsized influence that data systems have over our lives. Thus, when we choose to adopt a new device or app (or, worse, when we don't have a choice), we give up the power to choose "how other people attribute to you certain kinds of personhood or identity." Those kinds of judgments, she says, are hardwired into the tech, and once the tech itself is hardwired into our bodies and lives, we've lost the ability to define ourselves according to our own value systems.

Forlano believes that her nightmarish experiences using a smart insulin pump are a troubling sign of what's in store as algo-vision becomes a common dimension of lived experience for nearly everyone across the globe. She hopes that technology designers and regulators alike, as well as everyday technology users, will heed this warning. "People that are using these medical

I'm a PC.   I'm a Mac.

Figure 6.2
"I'm a Mac, I'm a PC" Apple ad campaign from 2006.

devices, for an existential reason, are kind of the canaries in the coal mine for a lot of these questions," she told us. "Because we have to do it, just in order to get through the day."

ALGO-VISION AS SUPERPOWER

Algo-vision might be an unavoidable consequence of living in a data-saturated, hyperconnected social environment, but it doesn't mean we're powerless in the face of technological dictates. As we mentioned above, people have a way of gaming the systems of surveillance and control that surround them, and under the right circumstances, algo-vision can be a very useful tool of resistance, whether you're raging against the machine or just trying to carve out a little privacy in your personal life.

One of the most basic ways in which people push back against the power of algorithms is just by trying to understand them. Technology companies love to show us how much they know about us—think about Spotify's

Wrapped feature, which sends each user an annual customized description of their musical tastes and activities over the past year, or Gmail's autocomplete function, which guesses what you plan to write next, based in part on what you've written before. These can be delightful tools of self-discovery ("Gee, I didn't know I listened to so much Simone Dinnerstein!"), public forms of self-promotion ("Hey everyone, look at how eclectic my tastes are!"), or horrifying portraits of our subconscious minds ("Why does autocomplete think I want to write a racial slur?").

But what technology companies don't typically do is tell us exactly how they use our data to make their decisions. Nearly every algorithm that shapes our lives in networked society is a metaphorical black box whose inputs and outputs can (sometimes) be observed, but whose inner workings are proprietary trade secrets owned by big corporations. No single user can possibly figure out why a social media platform chose to block or boost their content, why their email app thinks they want to write a certain word, or why they've received a specific perk or coupon from an advertiser. Collectively, however, people often get together and share their individual experiences in order to reverse engineer a given algorithm's logic.

Ironically, this often happens via social media. For example, tech researchers Renkai Ma and Yubo Kou have documented how users on Reddit get together to share information after being blocked or demonetized by YouTube's content moderation systems, "gradually gaining and applying practical knowledge of algorithms . . . to speculate, make sense of, and reflect on algorithmic penalties, informing their behaviors of repairing and avoiding future punishments."[26] Along similar lines, researchers at think tank Data and Society have documented how online teens get together in the comments sections of TikTok videos to speculate about why the popular platform's content moderation algorithms sometimes lead to dangerous consequences, such as promoting videos by trans creators to transphobic viewers. Though these speculations by users aren't scientifically valid investigations, the collectively shared information contributes to what Data and Society researchers refer to as *algorithmic folklore*, allowing participants to develop common practices of resistance that may help to protect vulnerable users and limit the damage caused by an algorithm's unilateral decisions.[27]

Figure 6.3
Invisibility cloaks, computer vision dazzle, and Juggalo makeup. *Sources:* Tom Goldstein, Adam Harvey, Flickr user WaeDC.

Once people understand (or think they understand) what goes on inside an algorithm's black box, they can come up with some fairly creative ways to resist its power. One example that has received a lot of attention in recent years, especially from the tech-savvy internet geek crowd, is surveillance-blocking makeup and fashion. There are countless examples, including "invisibility cloaks" (technically, "adversarial attacks on object detectors") developed by computer scientists, creative projects like Adam Harvey's "computer vision dazzle" makeup concepts (featured in *Vogue* magazine), and even unintentionally successful techniques discovered via algorithmic folklore (such as the Juggalo makeup adopted by the most ardent fans of surrealist hip-hop collective Insane Clown Posse).[28] Beyond being functional, these antisurveillance techniques become aesthetic markers in their own right, reflecting the wearer's affinity with a larger community of resistance to surveillance, and inviting onlookers to consider their own visibility to surveillance networks.

Other examples of algo-visionary resistance come in the form of individual and collective attempts to mess with the algorithms and sensors themselves. Sometimes, these efforts are subtle and calculated, like a judo practitioner using an opponent's greater weight against them. And sometimes, they're just brute force, smashing tech repeatedly until it breaks.

Many users and communities have learned to feed misinformation selectively into data-processing algorithms in order to nudge them into making different decisions. Are cars speeding down the street where your children are trying to play? Report a traffic jam to navigation apps like Waze, and if enough people do the same, the app will reroute traffic away from your street.[29] Is your employer invasively monitoring your computer activity while you work from home? Buy a mouse jiggler device to convince the computer that you're hard at work while you're really out walking the dog or getting a snack.[30] Over time, the apps will adjust to account for these strategically false bits of user data, but then users will eventually adjust to those adjustments.

Other communities have pooled their resources to try to pry open the black boxes themselves, jury-rigging techniques for making proprietary algorithms more transparent and accountable. For instance, ride-hailing app Uber has been caught (and has formally acknowledged) underpaying its drivers for years via a variety of methods, including undercalculating the distance of individual rides. That's why software engineer Armin Samii created a browser extension, UberCheats, which automatically compared the ride distances according to Uber against those calculated independently by Google Maps, giving drivers the opportunity to find out exactly how much they were being shortchanged by the company. Uber subsequently forced Google to remove UberCheats from Chrome, based on a somewhat dubious claim of trademark confusion, but by that point, Samii's app had already helped to bring about broader awareness and deeper scrutiny of Uber's exploitative practices.[31]

You may be thinking that invisibility cloaks, fake traffic jams, and distance-checking browser extensions all sound really cool and useful, but they also sound like a lot of extra hassle. If so, you've got a point. Even these successful uses of algo-vision as a resistance superpower are still the unfortunate consequences of pervasive data surveillance—an extra tax imposed on everyday people by virtue of living in a data-rich social environment. Just as we wouldn't need heroes like Spider-Man if New York City weren't so riddled with muggers and aspiring supervillains, we wouldn't need apps like UberCheats if our professional lives weren't so riddled with unaccountable sensors and algorithms. With great power (to surveil) comes great responsibility (to resist).

Thankfully, another consequence of algo-vision is that some companies are getting smarter about the secret lives of the data they capture, store, and create, taking steps to help people mitigate their most damaging social consequences. For centuries, for instance, many newspapers routinely covered local crime stories, publishing the names of suspected criminals provided by the police. While this might present some embarrassment or inconvenience for the accused (especially the innocent ones!), it was usually a temporary problem; today's newspaper is tomorrow's fish wrap. But, in the age of Google, old news stories are easily searched and retrieved by anyone, anywhere, and being named as a suspect in a crime can prevent someone from landing a job or signing a lease, decades after a court has established their innocence. That's why, in 2021, the Associated Press announced it will no longer name suspects in minor crimes, and the *Boston Globe* instituted a new policy allowing people to request that their names be removed from archived news stories.[32]

Even newer companies, born during the internet era, have had to change their policies in recent years to protect customers from the unforeseen consequences of their data collection practices—a rare step in a business environment characterized by the Silicon Valley philosophy of moving fast and breaking things. Social media giant Snapchat eliminated a feature called "speed filter" in 2021 not because it was unpopular, but because it was too popular. The feature, which allowed users to record their speed and share it with friends, was credibly blamed for encouraging reckless driving (which perhaps wasn't an entirely unpredictable outcome in today's "doing it for the likes" social media environment).[33] And in 2022, Amazon decided to stop including details about online purchases in the confirmation emails it sent customers, not because the information was irrelevant, but because it was too relevant—to the company's competitors. Google had been automatically scanning Gmail users' inboxes to profile their purchasing habits based on these automated messages, and Amazon didn't want to keep sharing such valuable information for free.[34]

At the same time, many other companies have been using their algo-vision to achieve less altruistic ends. In recent years, a new set of techniques has cropped up around using small, sometimes imperceptible changes in the design of an app or website to nudge users into making decisions they wouldn't have made otherwise, such as clicking on an ad, or purchasing a subscription. Companies with millions of online users have the opportunity to A/B test, trying out one version of a design on some people and another version on others, then instantly adopting the more successful of the two. Leveraging these insights, they have become adept at making the minimum necessary changes to their interfaces in order to effect the maximum possible changes in users' behavior, while still maintaining the illusion that users are in full control of their own decisions. Sometimes, these techniques are so subtle you'd never know there was a change—for example, diminishing the size of a button to close a pop-up ad by a small fraction. And sometimes, the techniques are a little more clever, such as overlaying an advertisement with the image of a single hair or a small insect, so that users will try to swipe their screens clean, and in so doing, accidentally click on the ad itself.

These secret modes of persuasion are appropriately referred to as *dark patterns*, and thankfully they've started to come to the attention of regulatory agencies, with the US Federal Trade Commission issuing an enforcement policy against them in 2021, and the European Data Protection Board issuing formal guidelines for designers and consumers to avoid dark patterns the following year.[35] Yet, because marketing technology tends to move at a faster pace than regulatory policy, there will probably always be a gap between the latest forms of algo-visual persuasion and regulators' ability to identify them and hold their designers accountable. In the words of FTC commissioner Rohit Chopra, trying to stop dark patterns without a comprehensive understanding of the problem, let alone a regulatory toolkit that is up to the task, can be like "playing whac-a-mole," always reacting to the latest threat and never addressing the underlying challenge.[36]

FROM SPIDER-MAN TO IRON MAN?

The QS movement, social media dysmorphia, and algo-vision all relate to the same basic set of social conditions. Those of us who are surrounded by data collection systems and subject to the algorithms that use them to make decisions about our lives have responded by doing what human beings always do in the face of powerful institutions: we try to figure out how these systems work, and then turn them to our own advantage by any means necessary. We collect and analyze data from our personal devices so we can track ourselves with the same precision that corporations and states employ to track us. We change our clothes, makeup, and bodies to fool algorithms or to bend them to our own purposes. We share information with other workers, citizens, and consumers, and try to pry open black boxes to access the secrets inside. By thinking like a data collector, we learn to flex a new set of superpowers, for good or for evil.

But these technosocial conditions are short-lived. If regulators are always two steps behind emerging developments in data and society, we also tend to be a step behind at best. Thanks to Moore's Law, industrial capitalism, and the lure of political power, new systems of data collection, analysis, and control will inevitably continue to warp our social fabric and require us to keep altering our lives and identities accordingly.

When we interviewed technology developers, they described a range of different visions for the future, from near-term developments like augmented and virtual reality to long-term, science-fictional stuff like the singularity, a melding of human and machine consciousness envisioned by futurist Ray Kurzweil as an almost spiritual transcendence of the human condition.[37] But one thing that many technologists tend to agree on is that the current trend points to the development of what is often referred to in the industry as the "Iron Man suit"—a suite of wearable, implantable, and environmental devices that will create a permanent data bubble around a given individual, augmenting their senses, extending their sphere of influence, and assisting in both physical and intellectual tasks throughout every aspect of their lives. Those of us with smartwatches, smartphones, networked medical implants, or smart homes are already part of the way there.

Nothing about the future is guaranteed, and the only thing we can be certain of is that our imaginations will invariably fall short of whatever reality can serve up. Yet, if so many technology developers and scholars believe we're on track to spend our lives in a cyborgian, transhuman, data-augmented bubble, we need to start thinking about how this so-called Iron Man suit might affect our inner lives, at what cost, and to whose benefit.

Thus far, algo-vision has been a superpower precisely because the data systems that pervade our lives and influence our decisions are piecemeal, obscure, and incomplete, and that leaves a lot of space for communal guesswork and resistance. But what kinds of space will be left for such things if each of us is fitted with an Iron Man suit at birth, and we live in its totalizing, protective bubble for the rest of our days? Unless we heed the warnings of forward-thinking tech critics like Laura Forlano and bring more diverse voices and perspectives into the design of data systems and assistive technologies, we may find ourselves locked into bodies, identities, and subjectivities that were developed not in the interest of empowering us but in the interest of profit or political control. Furthermore, if we continue to relate to technology through the solipsistic lens of an individual user, rather than through the shared strength of our communities and cultures, we risk losing the fundamental autonomy and freedom of choice that are enshrined as human rights in modern democratic society. The stakes could not be higher.

In other words, the choice before us isn't between having technology-assisted superpowers and living as mere mortals; it's about whether we can adopt new technologies while preserving the far greater strength that comes from living in a community of equals, sharing power, resources, and information fairly and freely. "Basically," Forlano says, "it's about human agency."

7 ALL THE WORLD'S A STACK

The twentieth-century Argentinian author Jorge Luis Borges has become something of a patron saint to twenty-first-century information scientists and scholars of data and society. Though he was writing at the dawn of digital computing, Borges seems to have had an almost supernatural prescience about how networked information flows might reshape the social landscape. Stories like "The Library of Babel,"[1] "The Book of Sand,"[2] and "The Aleph,"[3] though published close to eighty years ago, still resonate loudly in the tech world. Many of today's most popular narratives, from multiverse fiction to metaverse business plans, are clearly influenced by Borges's visionary work.

One of Borges's shortest stories is also one of his most enduring and influential. Titled "Del rigor en la ciencia" ("Of Exactitude in Science"), it's a one-paragraph tale about the map of an unnamed empire.[4] As imperial cartographers get better and better at documenting the world, the map itself becomes larger and larger, and increasingly detailed, until it reaches "the size of the Empire itself," with a one-to-one correspondence between the features drawn on the map and the geographical features they describe. At this point, of course, the map becomes useless, and the empire abandons it.

Critics and theorists have referenced this story for decades, and it's a well-known touchstone in information science and cultural studies.[5] It's the perfect illustration (so to speak) of a central dilemma in language and culture: in order for a map or symbol to be useful, it must have some relationship to the object or idea it's representing. If there's too little correspondence, then it will seem arbitrary and have little use. Imagine a paper map, for instance, in which the earth's continents are just green squares in a sea of blue. No traveler would ever need it, because it would be impossible to plot a route from point A to point B. On the other hand, if there's too much correspondence between a symbol and the object it refers to, then the symbol becomes redundant. You could never use Borges's empire-sized map because it wouldn't fit in your pocket—or even in your city. And even if you managed to fold it up really tightly, you'd have just as much trouble searching for a feature on the map as you would searching for a spot in the real world.

Representing the physical world symbolically, therefore, is a challenge of scale. In theory, there's an optimal balance between too much detail and too little, but it's challenging to find that balance with any precision, and naturally it differs from map to map, reader to reader, and context to context. If you've ever tried to build a piece of furniture at home from a prefabricated kit, for instance, you might have found yourself wishing for a bit more detail in the instruction manual. On the other hand, like Borges's imperial cartographers, real-life designers often don't realize they've gone overboard until it's too late; capturing and rendering information about the world around us has become so easy that the temptation to cram more of it into every corner of every interface can be overwhelming. Yet those of us living in networked society now face a fresh version of the dilemma, which not even Borges predicted: what happens when the level of detail available on our maps *exceeds* what we perceive in the world around us?

At first blush, this might sound like an impossible scenario; how could a map include more details than the terrain it describes? Once you include every possible feature, wouldn't the map be complete? Not if you take the secret life of data into account, because, as we've discussed, data has a way of begetting more data. Today's maps represent, in addition to the physical terrain, the social terrain, the cultural terrain, and the informatic terrain, all of which interact in ways that generate yet more information.

On a consumer-grade app like Google Maps, for instance, you can zoom in to nearly any spot on earth so closely that you can make out individual pedestrians, cars, and storefronts. But you can also overlay those photographic data with geotagged metadata such as the names of streets and landmarks; elevation data, including three-dimensional renderings of buildings; real-time traffic data; customized route-planning data; street-level photos and videos from a pedestrian's point of view; business listings and ratings; and thousands of other layers of information provided by Google and third parties, including by individual users.

To make the situation even more complex, the physical world itself now contains so many computer sensors and networking nodes that the terrain increasingly functions as part of the map by feeding data directly into various layers within apps and integrating networked technologies such as QR codes and Wi-Fi hot spots into the physical environment itself. As a result, the distinction between map and terrain has blurred to the point that, for many of us traversing physical and digital space, it's difficult to say where one starts and the other begins. With the dawn of new mixed-reality technology, such as computer-enhanced eyewear that analyzes the wearer's physical surroundings and overlays data and virtual objects in their field of vision, this blurring between physical and digital space seems almost certain to accelerate in the coming years.[6]

One way to frame this state of affairs is that data systems have thoroughly saturated our physical environment. But it makes just as much sense to view the situation from the opposite (algo-visual) perspective: from the vantage point of a data system, the natural world has become simply one more layer in the stack. And that means that the very concept of a map has become inadequate to understand the relationship between information space and physical space.

A map is purely derivative. Its job is to represent something larger, fixed, and external, in order to view it as a whole, "from above" (or, as cultural theorist Donna Haraway famously put it, using the "God trick").[7] But adding the physical world to the stack means that the terrain is no longer just something we attempt to see so we can navigate it better. It's now become something we attempt to *calculate* so we can understand (and therefore control) it better.

Environmental objects have thus become data objects. That means they're not only quantifiable; they're also accessible to statistical analysis, and may be put into mathematical relationships with other data. This has created a field day for statisticians, who now comb through what may seem like widely divergent datasets looking for correlations. Most of what they find are mathematical coincidences that have no underlying cause. To name just one example, data scientist Daria Kostyannikova has calculated a correlation of nearly 99 percent between US yogurt consumption volume and carbon dioxide levels in China from 2000 to 2016—a statistical relationship that she considers definitively spurious.[8]

Sometimes, however, there's a more likely causal relationship to be established. In 2021, for instance, psychologists at the University of Virginia published a study showing that "the number of lynching victims in a [US] county is a positive and significant predictor of Confederate memorializations in that county." The researchers would never have been able to demonstrate this relationship without the ability to integrate two completely different datasets: a "county-level lynching registry" created by a nonprofit organization called the Equal Justice Initiative, and a detailed catalog of more than two thousand Confederate memorials compiled by civil rights organization Southern Poverty Law Center.[9]

But even revelations such as those in the UVA study pale when we consider the broader implications of integrating our physical environment into the stack. All of the social consequences of the secret life of data that we've discussed throughout this book, from algorithmic bias to state and commercial surveillance to algo-vision, are now part and parcel of our relationship to the physical world we inhabit, whether we live in a pristine wilderness or a bustling megalopolis. This stackification of the natural world drastically expands what urban planning and environmental science scholars refer to as the *viewshed*—the scope of geographical space that is visible, and therefore knowable, from any given vantage point. This expansion gives us both an unprecedented degree of power and an unprecedented level of responsibility to understand how our individual and collective decisions shape the world around us, and vice versa.

CROWDSOURCING STEWARDSHIP

"Catch a falling star."

From John Donne's classic 1633 poem to Perry Como's hit 1957 song, the idea of tracing the bright trajectory of a meteorite from our upper atmosphere to the exact spot on earth where it lands has been used as a metaphor for impossible hopes and dreams. In Donne's poem, the idea is just as unattainable as learning to hear the songs of mermaids, or impregnating a mandrake root. And even by Como's day, the idea was still pure speculation; an observed meteorite was recovered for the first time in 1959, two years after his song originally hit the airwaves. Yet today, thanks to a handful of planetary scientists and a loosely cobbled-together network of more than ten thousand individual smartphone users around the globe, this once-impossible dream has become a common reality.

It all began in 2005 with an initiative called the Desert Fireball Network based at Curtin University in Australia. Researchers there started with the premise that what was impossible for any single person to confirm on their own would become clearer if their observations could be assembled and integrated using an AI-based computational model, "triangulating trajectories from multiple viewpoints."[10] The researchers built a smartphone app called Fireballs in the Sky and encouraged its thousands of users to point their cameras at celestial locations associated with meteor showers. Whenever these citizen scientists spot a shooting star, the app collects geolocation information and other metadata about the video and feeds them into the AI model, which combines these observations to predict whether and where the meteorite has landed.

The project, which was expanded through a series of institutional partnerships to encompass a Global Fireball Observatory (GFO), has been successful, with thousands of citizen scientists working in conjunction with a network of dedicated high-resolution cameras to track hundreds of fireball events each year. This has contributed to a record number of meteorite recoveries over the past few years—dozens, from Australia to Botswana to the United States.

The scientific benefits of this project are considerable, from giving geologists deeper insight into the origins of our planet and solar system to giving astronomers a better understanding of the threats posed by earth-crossing asteroids. But the GFO is just as remarkable for its process as for its product. We tend to think of smartphone cameras as selfie machines—solipsistic devices that risk isolating their users as easily as they may bring us together.

And typically, compiling and combining videos from multiple devices tends to be done in the interest of centralizing power and increasing social control, such as using Ring doorbell footage for police surveillance, as discussed in chapter 4. But the GFO is neither solipsistic nor paternalistic; instead, it provides us with a third model for how technology might be used collectively to everyone's benefit. By triangulating data from tens of thousands of users, the GFO sidesteps Haraway's "God-trick" and creates a map of the world that integrates multiple vantage points, exponentially increasing its accuracy while serving both the individual and shared needs of several different communities.

This model has become increasingly prevalent in the natural sciences. Project Noah, launched in 2010, archives and catalogs millions of wildlife photos taken by hundreds of thousands of citizen scientists around the globe with the aim of "crowdsourcing ecological data collection." It then uses these data for a variety of purposes, from research to education. Forest Watcher, an app developed by nonprofit environmental organization World Resources Institute, allows individual users and communities to document changes in their local ecologies, especially deforestation, and pools data from tens of thousands of individual users to inform research and policy. And FathomNet is an open-source database that has amassed more than twelve million photos and videos of the world's oceans from thousands of individual contributors and trained a machine learning algorithm on this dataset in order to help researchers more quickly identify species that appear in footage, and to better understand the impact of climate change on marine ecologies.

What unites all of these projects across land, sea, air, and space could best be described as *crowdsourced stewardship*: an ethic of shared responsibility to understand and protect the natural world, accomplished through capturing digital sensor data individually, sharing those data via online networks, and analyzing them for our collective benefit using AI and ML models. But good stewardship is about more than simply cataloging nature's many wonders (though that's probably the most fun part of it). New methods for capturing data about the world on a global scale have other functions as well, from exploring the past to planning for the future or policing the present.

LiDAR laser mapping technology has been used in recent years to create far more detailed topographical analyses of the earth's surface than was previously possible. This has already contributed to some major historical discoveries. For instance, archaeologists in the Mexican government and at the University of Arizona recently surveyed tens of thousands of square miles of southern Mexico using LiDAR and documented nearly five hundred previously unknown buried ceremonial sites from the Olmec and Mayan civilizations, which dominated the region for millennia. Some of these sites are so massive that they had previously been assumed to be hills and geographical features.[11]

In addition to revealing the existence of these structures, the LiDAR surveys have required historians to revise their understanding of ancient Mesoamerican civilizations. Not only were they broader in scope than previously believed, but the architectural features of the buried sites provide new insights into these societies' cosmological, architectural, and agricultural systems, their social institutions and cultural beliefs, and even specific ways in which the earlier Olmec civilization seems to have influenced the later Mayan one.

While archaeologists are using LiDAR scans of the earth to learn more about the past, astronauts are using the same technology on the moon to prepare for the future. Specifically, NASA has partnered with private tech companies to develop the Kinematic Navigation and Cartography Knapsack (KNaCK), a portable LiDAR system that will be worn as a backpack by humans on the lunar surface, allowing them to map the terrain with "an order of magnitude greater [precision] than conventional lunar topography maps and elevation models."[12] These maps will be a vital component of the Artemis Program, a plan to establish a more permanent human and robotic presence on the moon during the 2020s, in order to use it as a staging platform for future interplanetary voyages by human explorers.

Planetary-scale environmental sensors are also used for a variety of purposes related to the here and now. In 2022, an international team of environmental scientists published a paper in the journal *Science* that demonstrated how atmospheric monitoring satellites (specifically, a system called TRO-POMI) can be used to "quantify very large releases of atmospheric methane by oil and gas industry ultra-emitters" who account for about one-eighth of all global methane emissions. By identifying these highly polluting oil and gas businesses and "enforcing leak detection and repair strategies," the authors argue, environmental enforcers can prevent global warming by a degree equal to removing twenty million vehicles from the road for an entire year.[13]

In some cases, environmental data are being used to document actual criminal activity and hold the wrongdoers accountable, both in the wild and in densely populated urban environments. In 2021, for instance, a timber poacher was convicted of conspiracy, theft and other charges after accidentally starting a forest fire while logging hardwood illegally in the Olympia National Forest. For the first time ever, tree DNA was used to trace the accused poacher to the scene of the crime. The feat was only possible because the US Forest Service had assembled a database documenting the unique genetic profile of the poacher's target, the bigleaf maple, based on samples from over 1,100 individual trees.[14]

Along very different lines, a company called ShotSpotter offers what it calls a "precision policing platform," which it sells to law enforcement agencies in over 135 US cities, including New York, Chicago, and Boston. The foundation of the company's service is a network of acoustic sensors—always-on microphones, stationed in public locations. Whenever a gunshot is recorded by the microphones, the company triangulates its location based on multiple recordings of the event and alerts local police, as well as potentially pointing video surveillance cameras toward the geographic location associated with the sound. The company's website touts its successes, claiming credit for double-digit reductions in gun violence across a range of urban areas.[15]

A Motherboard investigation in 2021 challenged the validity of these claims, citing insufficient empirical evidence that ShotSpotter correctly identifies criminal shootings and reduces gun violence. Even worse, Motherboard found that "the company's analysts frequently modify alerts at the request of police departments—some of which appear to be grasping for evidence that supports their narrative of events."[16] Specifically, the article described ShotSpotter retroactively revising the location, timing, and number of gunshots reported by its system in order to bolster otherwise tenuous allegations brought by police against criminal suspects. ShotSpotter has disputed this story vehemently, writing on its website that "these untrue statements have been unfairly twisted to impersonate facts in the public dialogue."[17]

We have not reviewed the evidence underlying the claims by either ShotSpotter or Motherboard, and we can't confirm which party's assertions are accurate. However, we do consider it problematic that, unlike the GFO or Project Noah, the ShotSpotter system only triangulates data from sensors controlled by a single entity: itself. This gives ShotSpotter unilateral power to reshape data to fit a desired narrative or outcome. Therefore, whether the AI models built from these data are accurate or not, they lack the ethical and empirical validity that comes when multiple parties with divergent viewpoints (literally and figuratively) pool their information resources together to create a model of the world that's big and broad enough to reflect a multiplicity of perspectives. And, whatever the intentions or values of those who built and use the ShotSpotter system, its sole, unaccountable control over data collection and analysis undermines the platform's ability to serve the purposes of stewardship and the needs of a diverse community.

NEW DIRECTIONS IN ENVIRONMENTAL SURVEILLANCE

Many of the environmental data systems we've discussed rely on well-established and widely adopted sensor technology—digital cameras, microphones, satellite imaging, or DNA testing—along with AI and ML models to do their work. Some of them, such as KNaCK, use newer tech like LiDAR in novel environments (like the moon) to do something even more innovative. But a range of emerging technologies, techniques, and analytical models are likely to expand our capacity for environmental surveillance even further in the coming years.

One of the most potentially transformative technologies in this area is called *structured light*. We discussed this technique briefly in chapter 2 in the context of artist Kyle McDonald's experiments analyzing the illumination patterns on a ceiling to model the three-dimensional contents of an entire room. Computer vision engineers have made significant strides in this area in recent years, generating three-dimensional models of underwater environments based on imagery from a stationary camera posed above the water's surface, and three-dimensional models of open-air environments based on a single photo taken through a window spattered with raindrops, using the drops of water as miniature wide-angle lenses.[18]

What's interesting conceptually about these techniques is that obstacles to direct sight, such as water droplets, are reimagined as secondary information sources, providing additional data about the objects they obscure with the unique ways in which they warp and block light. However, what's driving most of the research in this field isn't mere theoretical curiosity—it's strategic interest from military and industry-based funders. Structured light has the potential to help self-driving vehicles, autonomous battle drones, and ballistic weaponry get a better, faster, more accurate lay of the land than many existing forms of environmental surveillance, despite many of the factors that typically impede these efforts, from inclement weather to battlefield smokescreens.

Another example of cutting-edge technology accelerating the stackification of our environment is *mixed reality* (MR), which we also alluded to briefly above. True to its name, MR aims to mix together elements of geographic space and dataspace into one (ideally) seamless experience for a user. This is typically a three-stage process: an MR platform absorbs information about the physical world, such as audio, video, or architectural sensor data; combines it with information from databases and AI models; and then presents the end user with a layered or augmented version of the reality they normally experience with their eyes and ears (there are fascinating ways to digitize and render information corresponding to our other senses, such as touch, smell, and taste, but they are not nearly as well developed or widely adopted at this point in time).

As we write this chapter there are countless startups, tech titans, and outright fraudsters operating in the MR space, and, for well over a decade, mixed reality has been one of those technologies that seems perpetually on the verge of taking off. (You may remember the accelerated cycle of hoopla, backlash, and disappointment that attended the announcement of the Google Glass MR platform in the early 2010s.) However, despite seeming eternally not ready for prime time, there are several reasons that we believe MR in some form will soon find wide adoption.

For one thing, it's just so darn useful. Science-fictional movies, television shows, video games, books, and comics are rife with examples of near-future scenarios in which people adorned with MR devices use them for professional, personal, creative, criminal, military, and pedagogical purposes, and audience members apparently have an easy time imagining such devices playing these roles in their own lives.[19] This is an important hurdle for any technology to overcome in order to achieve wide adoption (even if many of these speculative tales range from the cautionary to the dystopian). So demand from both consumers and businesses is likely to skyrocket once MR becomes a full-fledged, easy-to-use suite of technologies.

The supply side of the equation shows just as much promise for the future of MR as the demand side. Big tech companies like Microsoft, Magic Leap, Meta, and Apple have already invested tens of billions of dollars in MR acquisitions, research, and development over the past decade, and market intelligence firm Valuates projected in 2022 that MR would reach an aggregate market value of nearly half a trillion dollars by 2030.[20] That's a big number, but one that makes sense considering what a large bet these companies have already made on the tech. This means that commercial versions of MR technology are likely to hit the market in the near future. Maybe by the time you read this book, the killer app for MR will already be capturing imaginations, spawning imitators, and spurring a lot of hand-wringing among critics of technology and society.

Current projects underway at some of these companies provide an interesting glimpse into how MR might further stack-ify the world around us. Meta, for instance, has committed its future strategy heavily to what it calls the metaverse, which is a catchall term it uses for its MR offerings.[21] Part of this suite of technologies is a prototype of "smart glasses" called Project Aria, which, according to Meta's website, aim to add "a 3D layer of useful, contextually-relevant, and meaningful information on top of the physical world" for the people who wear them.[22]

Thousands of Project Aria beta testers have been using the devices since 2020, walking around public spaces in the United States and Singapore in order to collect both sensory and biometric data to populate LiveMaps, a pilot project at the company that aims to generate a real-time, three-dimensional, crowdsourced, immersive model of the world overlaid with various kinds of contextual information. Ultimately, Meta hopes that Project Aria and LiveMaps will serve as flip sides of the same coin: billions of users will capture a continuous stream of data about their physical environs and biometric responses, while navigating a hybrid of dataspace and geographic space that integrates the sensor data of billions of other users. To put it another way, the company that renamed itself after Stephenson's cautionary parable about MR is building a twenty-first-century version of the product in Borges' cautionary parable about maps.

It's a pretty huge gamble, and lots could go wrong. The project could be met with consumer apathy, organized techlash, or industry-wide failure to rally around a common set of standards, protocols, and practices. But there's also a good chance it will pay off like gangbusters in the not-so-distant future and have a profound impact on the way millions or even billions of us socialize and interact. At the very least, the Project Aria prototype offers a glimpse into how the secret life of data may play out as MR becomes increasingly commonplace.

COLONIZING THE COMMONS?

On the Project Aria website, Meta highlights the steps it says it has taken to develop the project ethically: securing personal data, alerting pedestrians to the presence of a surveillance device, and "asking people of diverse genders, heights, body types, ethnicities, and abilities to participate in the research program" to minimize algorithmic and design bias.[23] Yet, as tech scholars Sally Applin and Catherine Flick pointed out in a 2021 article in the *Journal of Responsible Technology*, this may not be nearly enough to prevent the "potentially serious consequences" of Project Aria and products like it.[24]

The chief problem, according to Applin and Flick, is that commercial MR projects like Project Aria privatize—or "colonize"—public space, turning information derived from what belongs to everyone and no one into a private commodity controlled by a single corporation. In addition to commandeering our shared resources ("the Commons," in their words), this disempowers people in at least two ways. First, public space isn't just populated by architectural features such as buildings and trees; it also includes *the public*. Our bodies and identities are caught in the dragnet of Project Aria's omnivorous surveillance process, and continuous, real-time data about us become a corporation's private property every time we step out of our homes.[25] Second, even when MR products allow pedestrians to opt out of data collection (for Project Aria, members of the public can scan a QR code worn on a lanyard by a beta tester), it becomes *our* burden to register our disapproval, rather than *their* burden to gain our approval. This problem becomes geometrically more challenging when multiple commercial competitors launch their own equivalents of Project Aria. Affirming basic privacy in the physical world thus becomes a never-ending task that Applin and Flick call "akin to unsubscribing from junk emails."

The threat of colonization outlined by Applin and Flick is not unique to Meta, or even to MR. In fact, the metaphor of a privatized commons was frequently used by early advocates for a free and open internet. One of the most vocal proponents of this perspective was Lawrence Lessig, the celebrated law professor who also cofounded the open licensing collective Creative Commons, an organization whose very name reflects this framing.[26] In the years since, as networking technology has spread out into our physical environment, blurring the lines between online and offline and stack-ifying the world around us, the privatization of the commons by data-hungry businesses has become less of a metaphor about our use of the internet as a virtual public space, and more of a reality about the ways in which actual public space is altered by the ubiquitous presence of networked commercial algorithms.

The examples are too numerous to list. Some are mostly invisible to those of us who inhabit the commons physically. Street View, the Google Maps feature that allows computer users to see panoramas from pedestrians' vantage points at specific geographical locations, relies on a fleet of cars mounted with multidirectional camera rigs in order to collect its photographic data. But, as it turns out, that's not all the cars collect; Google first admitted in 2010 that it had "failed badly" by "accidentally" collecting other data from Wi-Fi hotspots its cars had passed in over thirty countries. After a decade-long class action lawsuit, the story was revealed to be more sinister: Google engineers had intentionally outfitted the cars with systems that would detect private Wi-Fi networks and collect personal data from them without permission, including individual internet users' emails and passwords.[27]

Other private incursions into public space by voracious data collectors are more obvious. In 2021 and 2022, several cities around the world, from New York to Honolulu to Singapore, began using semiautonomous robots to police urban environments. From Singapore's R2D2-esque Xavier to NYPD's quadrupedal Digidog, these devices have been tasked with collecting biometric data to scan for COVID-19 infections among homeless populations, infiltrating hostage situations, and identifying people engaged in "undesirable social behaviors" like smoking and congregating in groups of more than five.[28]

Seeing these data collection devices deployed in the commons tends to make people uncomfortable. Most urban environments have been ringed with CCTV cameras for decades now, and though they spurred a wave of protest when they were first installed, they have largely receded into the background for most pedestrians. But the personification of data collection in the form of a robotic dog or police officer still looks a bit too much like dystopian science fiction for most people in the 2020s. The NYPD had to retire Xavier early due to continuing pressure from both residents and lawmakers, who, in addition to calling the robot "creepy" and "alienating," raised concerns about potential bias against racial minorities and other vulnerable populations.[29]

These concerns about bias are justified. It's not merely that police departments like the NYPD have a long and well-documented history of surveilling, harassing, arresting, and assaulting people of color much more frequently than white residents, or even that disproportionately more surveillance technologies tend to be installed in neighborhoods populated by Black and brown families.[30] The concern is that, when data surveillance is added to the mix, these practices intersect with the algorithmic biases embedded in the platforms themselves, creating a feedback loop between society and technology that puts racial minorities permanently in the crosshairs.

There is already ample evidence of this dynamic in the case of CCTV cameras. As a 2022 investigative news story at ProPublica showed, ostensibly "race neutral" traffic cameras in Chicago led to twice as many tickets for drivers in majority Black and Hispanic neighborhoods as for residents of majority white areas.[31] These tickets, in turn, contributed to disproportionately greater financial hardship, more vehicle impoundments, and more interactions with the judicial system, feeding back into the social conditions that are used to justify greater surveillance of Black and brown communities to begin with, in a nightmarish, self-sustaining cycle. Given this legacy, is it any wonder that New Yorkers would greet the arrival of the NYPD's robotic dogs with a resounding Bronx cheer?

THE MÖBIUS STACK

Imagine you're an office worker back in the 1990s, hunched over your tan computer keyboard, in your tan cubicle, tapping numbers into a spreadsheet flickering on your tan CRT monitor. Microsoft Excel, the "productivity app" you're using, is just as bland as your office surroundings, with endless rows and columns of identically sized gray boxes, waiting for you to input data. The aesthetic of blandness isn't accidental—it's a strategy designed to keep you focused on the task at hand, and to discourage any flights of fancy.

After a quick glance back over your shoulder to make sure nobody's watching, you open a new workbook in Excel and press the F5 key on your chunky keyboard, type the text string "L97:X97," hit Enter and then Tab, and for your final coup de grâce, press Control and Shift together while selecting the Chart Wizard tool with your mouse. All of a sudden, the drab world falls away, and you find yourself soaring over an undulating purple landscape toward a distant horizon painted in yellows, oranges, reds, and blues. Using your mouse and the arrow keys on your keyboard, you roll, climb, and dive, exploring this new terrain from your virtual spaceship. You hear footsteps approaching your cubicle and hastily hit the Escape key. The simulated planet disappears, instantly replaced with rows of identical gray boxes. You've returned to the world of drab, but now you have a secret escape hatch: you've found an Easter egg.

If you're reading this book, you've likely heard of Easter eggs, and the chances are good that you've actually found and used one. Easter eggs are bits of media or software hidden inside other bits of media or software, and they're used for all kinds of purposes, from whimsy to marketing to political protest. They can even be exploited for criminal and cyberwarfare purposes; in these cases, they're typically referred to as *malware*.[32] But, for the purposes of our current discussion, what's most interesting about Easter eggs is that they show us how convoluted and weird the stack can be, and how its unexpected twists and turns can serve as conduits for the secret life of data.

When engineers and tech researchers talk about the stack, we tend to treat the metaphor literally, as though each system were placed contiguously above or below another one—firmware stacked on top of hardware, operating systems stacked on top of firmware, and so on, all the way "up" to the user interface. More sophisticated diagrams sometimes show the stack as a branching tree structure in which, perhaps, several different kinds of operating systems might be stacked on top of a single set of hardware and firmware. But even these models have a clear "up" and "down," and nice, clean lines demonstrate where one level of the stack ends and the next one begins.

This linear metaphor is very useful to engineers, who need its clarity and simplicity to organize the complex process of developing new hardware and software. But in many cases, the actual relationship between information technologies is anything but linear. The stack is really more of a fractal: layers may contain other layers, and strings of layers can look more like an infinitely recursive Möbius strip than a straight line. To make things even more convoluted, every transfer of information from one layer to another constitutes an act of *interpretation*—in fact, engineers sometimes refer to this process of moving information up and down the stack as *translation*. And, as cultural studies scholars have been telling us for the better part of a century, every act of translation is also an exercise of power: one potential interpretation must be privileged over countless others, and whoever decides how X translates to Y gets to choose what kinds of meanings are created, and what kinds are not.

In the case of a computing system, the moment of translation also introduces an element of chance and unpredictability. That's because one layer of the stack may "think" of the data in another layer not as a set of static

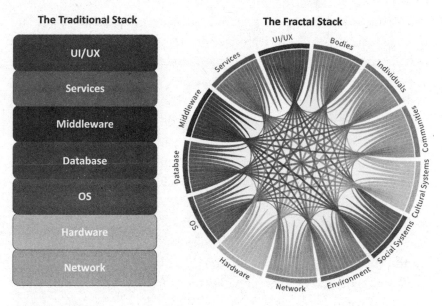

Figure 7.1

A traditional rendition of the stack versus the fractal stack. Fractal image generated using flourish.studio.

numerical values, but as an algorithm or set of instructions, with any number of possible outcomes. This is both a feature and a bug: it gives programmers immense flexibility and power in designing complex systems, but it also creates a lot of opportunities for error and exploitation. As we've discussed before, the metaphor of the Trojan horse is an appropriate way to think about the hidden agendas that may be lurking in a seemingly innocuous piece of information.[33]

There are many different varieties of Easter egg, each of which warps the stack in its own way, providing new outlets for data's secret lives, and sometimes revealing the hidden circuitry of social power in the process. The Excel 97 flight simulator, for instance, wasn't merely a bit of tomfoolery on the part of the coders, or a special feature dreamed up by Microsoft's marketing team. In fact, it wasn't even discovered until at least 1999, two years after the software was released.

According to an interview with Ed Fries, a longtime Microsoft executive who worked on Excel before leading the launch of the company's XBOX video game console, there was a "culture of Easter eggs" at Microsoft during its early years.[34] Nearly every productivity software product had one, but only the specific teams working on a given product knew about them, while upper management at the company maintained "plausible deniability" about their existence.

The revelation of the flight sim Easter egg, therefore, also revealed certain things about the company itself: First, contrary to its buttoned-down image, Microsoft had what Fries called a "'Hacker' kind of culture," marked by a playful relationship with technology. Second, business units within the company worked in relative isolation and autonomy from one another, and the audacity of individual Easter eggs was a source of competitive pride between parallel business units ("Look what we pulled off! Bet you can't one-up us!"). Third, Microsoft's flagship software products had, by the 1990s, already become so code-heavy with multiple iterations and updates (what engineers sometimes refer to as *bloatware*) that *you could stick an entire flight simulator inside of a spreadsheet app* and nobody would notice for two full years.

The Excel 97 flight sim might have been technically unsanctioned and tacitly critical of Microsoft as a company, but as hacks go it was fairly benign. Nobody's computer was destroyed, everybody had fun, and the company gained about as much cred for tolerating these hijinks as it lost for allowing bloatware stuffed full of extraneous code. In other cases, however, Easter eggs can wreak a lot of havoc, and their damaging consequences can extend far beyond even the malicious intentions of their creators.

This was exactly the case with the infamous Stuxnet worm, which was first discovered in 2010, infecting and disrupting industrial control systems operated by Siemens. Cybersecurity researchers found that Stuxnet piggybacked on the Windows operating system, taking advantage of a previously unrecognized vulnerability in the code to embed itself in a computer system and spread to others.[35] As with the Excel flight sim, the discovery of this software package inside of another software package led to additional revelations about the code's origins, in this case with major geopolitical repercussions. Researchers ultimately traced Stuxnet back to American and Israeli government intelligence communities' collaborative efforts to disrupt Iran's uranium enrichment program, which inadvertently revealed to the public both the Iranian nuclear program's existence and the coordinated cyberwarfare efforts of the United States and Israel.[36]

Easter eggs like the Excel flight sim and malware like Stuxnet are very crafty, but ultimately their success rests as much on human and organizational elements as on technological ones. Microsoft's permissive internal culture allowed an entire game to be smuggled inside of a productivity app and also allowed hackers from the intelligence community to exploit the software equivalent of an unlocked basement window in its operating system. In both cases, algorithms were hidden in one specific layer of the stack. If the software working in that layer had been more effectively inspected for vulnerabilities or extraneous code, the hacks would never have succeeded. So, in theory, these are preventable problems: if each layer of the stack is policed thoroughly on its own, the entire stack should be secure, like a chain in which each link's individual strength has been verified. But it turns out that's not completely accurate in practice, which is where things start to get really weird. Easter eggs and malware can also be encoded into the translation process *between* layers in a way that's not necessarily detectable in individual layers themselves.

For instance, computer scientists have known since the 1960s that even computers that are unconnected to any network can leak data in the form of electromagnetic radiation. If you're reading these words on an electronic device like an ebook or tablet, the resulting invisible radio waves that surround you at this very moment look different than they did when you read the previous page, or than they would if you were looking at a photo or video instead. Clever hackers can identify these different electromagnetic profiles and reverse engineer them to find out what is on your screen based solely on its unique radiation signature. The very poetic name for this kind of remote surveillance is *soft tempest*.

There are ways to defend a computer against a soft tempest attack, which include installing radiation-shielding technologies in rooms and devices, generating extra electromagnetic noise to drown out the targeted signals, and making sure that computer monitors face away from windows, which makes it much harder for someone outside a building to eavesdrop. But there are also "tempest viruses" that can be installed on a target device to make it leak *more* electromagnetic radiation than it would otherwise, making it easier to spy on even when the computer isn't connected to a network, or when various other protections are in place. And there are counter-counter-counter-espionage tactics like placing false information on a targeted device, or even installing a virus on the target device so that when hackers intercept its electromagnetic emanations and turn them back into digital information, they will inadvertently install the virus on their own devices in the process. And so on.

In other words, the information stack is far more topographically complex than the neat, linear diagrams that are usually used to illustrate it, and information can be hidden, transformed, extracted, and injected through its many obscure contours and channels. These convolutions take on a whole new (secret) life when living beings and global-scale networks are added to the stack and the lines between map and territory begin to blur.

There's a 1993 episode of *Star Trek: The Next Generation* entitled "The Chase" in which warring humanoid species from different parts of our galaxy simultaneously discover that their DNA contains shared and overlapping sequences corresponding to stellar coordinates. Convinced that their DNA also contains plans for a powerful weapon, members of each species race toward the designated coordinates, hoping that whoever arrives first will obtain the secret to galactic domination. Ultimately (spoiler alert) they all arrive at the same time and piece together the genetic fragments from each species to reveal not a weapon, but an ancient message from a common ancestor, imploring them to recognize their shared origins and to embrace one another as kin.

When this episode first aired, it seemed to many viewers (well, to us, at least) like a mind-blowingly cool but only distantly plausible scenario. After all, DNA had been discovered only forty years earlier, and using it to encode legible messages for future generations seemed about as likely as teleporting to a distant solar system. Yet, only a quarter of a century later, researchers from the University of Washington used DNA sequencing tools not only to encode a message into a living being, but to turn the message into malware. Specifically, these computer scientists encoded a strand of DNA that contained a kind of informatic poison pill: when the strand was analyzed on a computer using a common gene sequencing program, the resulting data would take the form of a self-executing computer virus that could take over the computer itself.[37] While this particular project was little more than a proof of concept, it demonstrates what is at stake now that biological and environmental systems are part of the information stack. There's no telling where Easter eggs and malware will be hiding, from microscopic molecules to planetary processes.

LIVING IN A CONNECTED WORLD

For better and for worse, the stackification of our bodies and our environment likely means that the physical world we inhabit will start to behave more and more like the internet, and our offline and online experiences will become less and less distinguishable from one another. Increasingly, the objects around us will collect data to feed into invisible repositories, and our interactions with those objects will be augmented with information synthesized from the same databases.

In practical terms, this means we can expect to do more stuff in real life that we're accustomed to experiencing on the web and in apps: hyperlinking between physical objects; searching the real world for specific items; instantly accessing metadata about the people and things around us; "sharing" our lived experiences with others via social media platforms; and, consequently, experiencing persistent state and commercial surveillance and suffering the consequences of algorithmic bias.[38]

Living in the stack also means that some of the unplanned and unforeseen challenges facing internet users will creep into our daily lives, as well. Many internet sites and services currently face the challenges of *link rot*, in which old URLs no longer point to functional web pages, and *content drift*, where links that were supposed to connect to one thing now connect to something else. This isn't an issue only for abandoned websites and derelict services; according to a recent analysis by Harvard Law professors, about a quarter of all links embedded in *New York Times* articles have been rendered unusable by link rot, and about one-eighth of the remaining links have seen significant content drift.[39]

On the other hand, there are millions of people around the world who would probably *prefer* to have information about them disappear from the internet, from embarrassing photos to poor work reviews to slanderous blog posts. The European Union and Argentina, to name two examples, currently give internet users a certain degree of control over this kind of content and have laws reflecting a "right to be forgotten" that can be enforced against internet platforms like Google and Facebook. Other countries, like the United States, have not enacted such laws, and internet users there have little power over their online reputations.

But even erasing undesirable data doesn't necessarily give people control over their online images and identities. That's because, as we discussed in chapter 2, in a data-rich environment, the *absence* of data can stick out like a sore thumb. Consider how damning the infamous eighteen minutes of blank tape were for Richard Nixon in the Watergate investigation. More recently, law professor Elizabeth Joh has raised concerns that when platforms like Google selectively erase location data to protect users who have visited sensitive locations like abortion clinics, the erasure itself could serve as a red flag to investigators in states that have recently outlawed abortion, leading to potential prosecution for these users.[40]

What can these challenges of retaining and expunging information on the web tell us about the issues we're likely to face as the lines between map and territory, online and offline, and planet and stack continue to blur? We can't know for certain, but we can be confident that the real-world equivalents of link rot and inexpungible records will present even greater hurdles than their originally digital equivalents. That's because as engineers and designers seek to create a seamless user experience connecting the physical with the digital, they are more likely to create new conduits for the secret life of data.

We interviewed cybersecurity researcher Vasilios Mavroudis of the Alan Turing Institute in London about this problem. As he explained to us, there's a fundamentally irreconcilable tension between the security of a system and its level of usefulness. Cybersecurity is a "party pooper," he said, because it can "kill usability." With all of the tools available to a twenty-first-century hacker, which Mavroudis believes will include forensic AI systems designed to analyze and identify vulnerabilities continuously, the only way to make a truly secure technological system is to reduce its features to the absolute minimum. The more layers you add to the stack, the more this tension grows, according to Mavroudis, because unforeseen threats are most likely to emerge "between the cracks of two systems," and "the scary thing is that . . . it goes down to the physical layer."

What's so scary about this? Well, for one thing, the physical layer is where we actually spend our time, working, playing, eating, and sleeping. For another, we still often tend to treat the physical world as a fail-safe against our digital problems. If your computer malfunctions, you can just switch it off and back on again and start from scratch. But there's no off switch for the world at large—at least, not one we'd ever want to toggle. And the more connections we build between the physical layer and the rest of the stack, the less able we'll be to disconnect or to exercise agency in the spaces between. Finally, as we'll discuss in the next chapter, our social and political systems are still built on the operating assumptions of a distinct and independent physical world. So the stackification of our bodies, environments, and institutions threatens to destabilize democracy and exacerbate social divides in ways that previous generations could never have anticipated, and to a degree that our current leaders seem unequipped to grasp, let alone fix.

8 DATA AND DEMOCRACY

Throughout this book, we've invoked a lot of fictional narratives, such as *Frankenstein, Blade Runner, 2001: A Space Odyssey, Smart House,* and "Del Rigor en la Ciencia." We've done this for a variety of reasons. First of all, we're fans. We love science fiction, speculative fiction, futuristic games and comics, and other stories that people tell when they try to get a handle on the overarching question of what might be. It's one of the things we bonded over when we first met as teenagers, and these kinds of stories have continued to fuel our conversations about technology and society ever since, right up through the writing of this book. But, more to the point, we refer to fiction frequently when discussing the real world because speculative narratives become touchstones for the popular imagination, encapsulating entire philosophies, fleshing out possible futures, and providing the public with a shared context to talk about the complex ways in which technologies continue to coevolve with social institutions. For instance, the burgeoning field of Afrofuturism seeks to reinvestigate old narratives and inspire new ones that center Afro-diasporic experiences and aesthetics in stories about the future of humanity.[1]

The power of narrative to frame the present and shape the future is one reason that we all must be very deliberate about the kinds of stories we tell, consume, and circulate. As those of us who research this field know all too well, today's pie-in-the-sky "what if" scenario can easily become tomorrow's reality—or, at least, can heavily influence it. As we've discussed in other publications, speculative fiction tends to cluster around a group of familiar *futuretypes*—common tropes invoked repeatedly in books, movies, and games, such as faster-than-light travel, sentient AI, or replicator technology.[2] Not only do these futuretypes reflect people's hopes and fears about the consequences of new scientific discoveries and inventions; they also influence the way that scientists, investors, and policy makers think and talk about the future, guiding the development and use of new technologies in a self-perpetuating cycle.

Consequently, even though we've asked the question, "What could go wrong?" many times throughout this book, we've mostly avoided invoking the subgenre of postapocalyptic fiction, which specializes in stories that explore scenarios in which something has gone very wrong. Part of our reluctance comes from not wanting to give such narratives too much oxygen—as we've mentioned, speculative fiction can be self-perpetuating, and if we invoke these tropes, we may legitimize them as possible futures. But more importantly, a lot of these stories just aren't very helpful to us, because they lack nuance and hope. Sure, a zombie plague *might* sweep through the vestiges of a desiccated, postnuclear hellscape on earth, but what's the point of talking about it if there's nothing useful to learn except where to point a shotgun?

The one exception we'll make to this rule is by invoking the sub-subgenre of post-postapocalyptic fiction. Simply speaking, these are stories about people learning to rebuild after a catastrophe. Books like Emily St. John Mandel's *Station Eleven*, David Brin's *The Postman*, and Octavia Butler's Parable book series show us how easily the fabric of civilization can unravel. But they also offer us a glimpse into how people might begin to stitch it back together through forging bonds of trust and mutual aid when all of our social institutions have failed. This futuretype is especially relevant here, in our final chapter, which addresses some of the threats to democratic norms and civil society that we now face around the globe, due in part to the secret life of data.

Democracy is, and has always been, enacted through algorithms. We don't mean computer programs, obviously, but rather algorithms in the broader sense of the word: clear operating guidelines based on quantifiable and reproducible protocols and procedures. Concepts like equal justice, fair representation, and an even balance of power are all admirable goals in theory, but the tough part of building and maintaining a democracy is figuring out how to put them into practice. In the United States, as in other democratic societies, we rely on countless interdependent methods of data collection and analysis, which are constantly adjusted and battled over, in order to inform our progress toward these goals. Whether drawing congressional districts, tallying electoral votes, or passing bills into law, our biggest decisions and most profound social changes often come down to the question of what counts, from a tiny hanging chad on a paper ballot to a giant corporation's taxable income. And, as evidenced by Donald Trump's plot to overturn the 2020 US presidential elections, who and what gets counted can also depend heavily on who gets to do the counting.

One of the main algorithmic principles that has informed democratic society since long before the internet age is the premise of equal access to information. As the political scientist Alexis de Tocqueville wrote in *Democracy in America*, his 1831 exploration of constitutional democracy in action, the invention of the postal mail system was a "great event" that "turned to the advantage of equality," because it brought "the same information to the door of the poor man's cottage and to the gate of the palace."[3] This kind of informational equality was essential to the American system, Tocqueville believed, because the US constitution "presupposes" that all citizens have access to a "variety of information" in order to exercise their constitutional rights.[4]

Brin's *The Postman*—perhaps inspired by Tocqueville's insight—takes this premise and turns it into a narrative of rebuilding democracy. The protagonist of the story, Gordon Krantz, is a former actor wandering in post-apocalyptic Oregon when he discovers a long-abandoned postal truck and postal worker's uniform. He initially dons the uniform for warmth, but soon he begins actually ferrying mail from community to community, giving their inhabitants a reason to believe in a "Restored United States."[5]

By the time Brin published his book in 1985, the idea of postal mail as a democratizing agent was more of a metaphor and less of a reality than it had been in Tocqueville's day. Television, radio, satellite, and even early computer networks were bringing much higher volumes of information to and from billions of people around the world at light speed. In the decades that followed, technologists and policymakers alike predicted that the "information superhighway" of interconnected personal computers and mobile phones would serve as a natural boon to democracy around the globe, a rising tide of data that would float all boats by bringing us closer together.

Yet, not long into the internet era, evidence began to suggest that the tide was not rising equally, and that gaps such as the "digital divide" between internet haves and have-nots were actually exacerbating long-standing social inequities. Even worse, it wasn't a problem that could be fixed simply by handing out free laptops or providing universal internet service. Along with our friend and colleague Arul Chib, we published research in 2008 that used agent-based computer models to show how giving new tools like email and mobile phones to everyone in an existing social network could actually have a *negative* impact on overall information equality, by giving privileged members of society faster access to salient information.[6] In the years since then, a significant amount of empirical research[7] has borne out these predictions.

Today, we're witnessing the consequences of these dynamics cascading to a geopolitical scale. At the time of writing, it's widely estimated that the majority of the world's population (roughly five billion people out of about eight billion) is connected to the internet, and the vast majority of people (roughly seven billion) have mobile phones. Yet, by many measures, the world is far less equal than it was before these technologies were adopted widely. According to analysis by the World Inequality Lab at the Paris School of Economics, the ratio of the average income of the top 10 percent of earners in a given country compared to the bottom 50 percent nearly doubled between 1980 and 2010 to a level not seen since the Victorian era, and it has plateaued at that level in the years since.[8]

At the same time, democratic norms and institutions are under assault in many of the world's largest constitutional democracies, including Brazil, India, the United Kingdom, Israel, and the United States. According to analysis by the International Institute for Democracy and Electoral Assistance (International IDEA), the "percentage of democracies with significant declines over 10 years" has skyrocketed during the information age, rising tenfold from about 5 percent in 1980 to 51 percent in 2020.[9]

In short, whatever hopes people may once have had that global computer networks and the apps and services built on them would play the role of Brin's and Tocqueville's postmen, bringing the world into greater harmony and equality through the global distribution of information, thus far the opposite appears to be the case. The vast data collection power of large sensor arrays like the IoT, the vast computational power of large computer networks, and the vast economic power of those wielding large information sets all multiply one another, conferring an incalculable degree of self-sustaining social power for the small handful of state and commercial institutions that control these resources. If democracy in practice resides in the minute calculations and institutional norms that maintain a healthy balance between different stakeholders, then it cannot continue to function when some stakeholders are able to calculate so much more effectively than others and violate long-standing norms with near impunity.

To complicate matters even further, many of the algorithmic tweaks that have undermined social equality and democratic institutions are computationally obscure or difficult for laypeople to understand and mitigate. Consider, for instance, the ways in which techniques like "cracking," "stacking," and "packing" are used to gerrymander congressional districts in the United States.[10] So, while democratic institutions may appear to be healthy and functioning on the surface, in reality they are being hollowed out from below, and the system is increasingly being gamed by those with the most data and processing power to bend policy to their own ends.

Democracy, in other words, has been hacked.

Yet, as we will argue at the end of this chapter, there's no such thing as absolute power, and the lessons of post-postapocalyptic narratives like *The Postman* may hold some promise for reestablishing democratic norms and sustaining institutions in the wake of these devastating disruptions. We'll end this chapter with a few rays of hope—we promise. But first, we'll need to delve into some dark, postapocalyptic territory, and take a closer look at what went wrong.

THE DIGITAL COLD WAR

If democracy has always had an algorithmic component, it's also true that war and diplomacy have always relied as much on informational tactics as on military ones. The history of espionage and counterespionage is long and colorful, from the "extensive system of spy-craft" employed by the French Reign of Terror to the birth of digital computing at Bletchley Park's Government Code and Cypher School during World War II, where the mathematician Alan Turing led a team of code breakers attempting to decipher encrypted Nazi communications.[11] During the Cold War that followed the Allied victory, spy craft and information technology became increasingly interconnected, both in the real world and in the popular imagination, as exemplified by Ian Fleming's fictional MI6 agent James Bond, with his endless array of gadgets and gizmos.

This is the world into which the modern internet was born. As early as the mid-1990s, preexisting mass surveillance networks such as ECHELON (a collaborative program of the Five Eyes nations: Australia, Canada, New Zealand, the United Kingdom, and the United States) had been expanded to include the interception of nearly all internet communications around the globe.[12] Meanwhile, many nations' intelligence agencies had launched their own internet surveillance platforms, such as Carnivore in the United States.[13] In the post-9/11 era, the so-called war on terror was used to justify further expansion of these surveillance practices and technologies, even as the rise of social networking platforms and widespread smartphone adoption continued to expand the scope of internet usage in everyday life. While periodic leaks about these top-secret government operations led to increased scrutiny by journalists and human rights advocates, general awareness of and interest in the scope of digital surveillance among the general population remained low.

The situation changed precipitously in June 2013, when an employee at a National Security Agency (NSA) subcontractor named Edward Snowden spilled some of the agency's most closely held secrets to the world. Using the access provided by his security clearance, Snowden downloaded millions of files from the NSA's servers related to global surveillance programs by the United States and its Five Eyes partners and leaked them to leading newspapers around the world. Among other things, these documents revealed existing programs enabling intelligence agencies to collect personal data about individuals from internet service providers and social media platforms, track individuals' locations in real time, and search anyone's email and chat histories. The leaked documents also revealed the NSA's plans to expand its scope of collection to encompass any available data from "anyone, anytime, anywhere."[14] Snowden's revelations had three more or less immediate consequences that altered the geopolitical landscape so radically that cybersecurity and internet governance experts now frequently refer to the "post-Snowden" era.

First, the documents demonstrated conclusively that the United States had been misleading its own citizens and closest allies about the scope of its data collection practices and ambitions. This discovery undermined both domestic support for the intelligence community and international goodwill for the United States (including willingness to participate in collaborative business ventures and political operations). Beginning in 2013, many US allies adopted a far more defensive posture when it came to internet governance, including building their own national internet infrastructures to avoid routing traffic through the United States and therefore feeding data to the NSA's digital dragnets.

Second, the leaks included what Glenn Greenwald (a journalist at *The Guardian* who was among the first to report on Snowden's findings) referred to at the time as "very sensitive, detailed blueprints of how the NSA does what they do."[15] In other words, Snowden didn't only reveal the existence of the NSA's far-reaching data surveillance practices; he gave detailed instructions to every other government and powerful institution in the world—including those of Russia, where he has resided permanently since the leak, and which granted him citizenship in 2022—about how to replicate them.

Third, due in large part to these consequences, Snowden's leak heralded a new era of cyberwarfare, cyberespionage, and cybersecurity in which nobody in business, government, civil society, or the general population could afford to ignore the significant risks that come with communicating over digital networks. Conversely, large governmental and commercial institutions could no longer afford to ignore the significant *opportunities* presented by data capture and analytical tools like those developed by the NSA. By demonstrating their feasibility, Snowden's revelations unleashed a new data arms race as a feature of modern statecraft and business competition.

In this new, post-Snowden era, digital networks and computer processors are no longer understood merely as accelerators or additions to age-old practices of warfare, espionage, and diplomacy, but as a battlefield unto themselves—in many cases, the first and most important site of contestation in any conflict.[16] As US Deputy Secretary of Defense Kathleen H. Hicks suggested in a 2022 memo to senior Pentagon leadership, this new state of affairs requires military institutions to fundamentally restructure and even reinvent themselves. In her words, "The Department of Defense must become a digital and artificial intelligence (AI)-enabled enterprise capable of operating at the speed and scale necessary to preserve military advantage."[17] This was not just rhetoric; the memo also appointed the nation's first chief digital and artificial intelligence officer, reporting directly to her office.

From a geopolitical standpoint, the digital arms race of the post-Snowden era has also ushered in what might best be described as a digital cold war (though there are some scholars and analysts who believe this framing does more harm than help).[18] However we choose to describe it, the reality at the time of writing this book is that all of the world's major political powers are currently engaged in ongoing antagonistic and opportunistic cyber operations, including but not limited to hacking government servers, surveilling government employees, conducting corporate espionage and intellectual property theft, authoring and distributing malware, and fomenting socially mediated disinformation and agitprop campaigns.[19] Even smaller nations like North Korea, Belarus, and Israel are punching far above their metaphorical weights, sometimes aligning with superpowers to accomplish collective goals (as with the Stuxnet worm), and sometimes operating independently to suit their own purposes (as when North Korea hacked Sony Pictures servers in apparent retaliation for distributing a film that satirized its leader, Kim Jong Un.)[20]

These state actors are frequently aided by commercial ones—popular consumer-facing companies such as online retailers and social media platforms as well as more obscure businesses like data brokers, internet infrastructure providers, and nongovernmental hacker collectives and troll farms. In authoritarian nations, there is little daylight between such organizations and the regimes they serve. In more democratic contexts, participation in state cyber operations is compelled via a combination of commercial contracting relationships, regulatory favor, and judicial intervention (such as warrants and plea deals). Either way, one cannot understand the digital cold war without recognizing that whatever lines may once have existed between commercial and state surveillance have now blurred to the point that they can no longer be viewed as discrete technologies or practices.[21]

If the lines between actors in the digital cold war are blurred, so are those between targets. The collateral damage frequently suffered by commercial entities and individual citizens in cyber warfare between rival states is far broader than in traditional warfare. This is due in large part to the technological idiosyncrasies of digital networks. Stuxnet may have been intended specifically to disrupt Iranian uranium enrichment facilities, for example, but it was a computer worm, and computer worms are made to self-replicate. So by the time it was detected, Stuxnet had infected industrial machines beyond Iran and as far afield as the United Kingdom, the United States, Russia, South Korea, Indonesia, Pakistan, and India.

In many cases, as journalists around the world have documented, civilians and businesses aren't merely victims of collateral damage from state-sponsored cyber operations, but the intended targets. China scrapes data from social media services like Facebook and TikTok to feed its military and intelligence services information about potential civilian targets in other nations.[22] Russia runs ongoing "brute force" attacks seeking password-protected access to hundreds of American and European organizations, from the energy sector to big media companies, looking for opportunities to surveil and disrupt operations.[23] The US Secret Service required travel-data brokers Sabre and Travelport to provide "complete and contemporaneous real-time account activity" of targeted foreign nationals for years under the questionable authority of a 1789 law allowing the government to request assistance in investigations from private entities.[24] The Saudi government used Pegasus spyware, developed by Israeli software firm NSO, to hack the phones of women journalists critical of their regime, and published their private and intimate photos on social media as part of assaults on their reputations.[25] These are a handful of examples; a comprehensive list would be longer than this book.

Although the Geneva Conventions make it a war crime to target civilians in a military conflict, there is no equivalent protection in international humanitarian law when it comes to cyberwarfare—even though such operations can easily cost lives (consider the consequences if a digital medical implant gets hacked, or if the entire electrical grid powering a hospital goes offline). Organizations like the International Committee of the Red Cross have published analyses arguing that cyberattacks should be covered by existing international humanitarian law, but nobody is currently prosecuting any such cases, and it's not clear that such efforts would result in enforcement with consequences.[26] Furthermore, as political scientist Nori Katagiri pointed out in a 2021 article in the *Journal of Cybersecurity*, international humanitarian law is limited in scope to state actors, but in many cases, private commercial or criminal enterprises are doing states' dirty work, creating what Katagiri calls a "legal void" in which "nonstate behaviors" are not policed.[27]

Finally, the digital cold war contributes to another kind of collateral damage that is not always recognized as such: namely, the erosion of democratic norms. The same technologies and programs used to hack and surveil foreign combatants or foreign nationals can be easily refocused on domestic targets, and the examples of such abuses are becoming increasingly frequent, even in formally democratic countries. Israeli police reportedly used the same Pegasus spyware employed by the Saudi government against foreign critics to hack the phones of its own citizens engaged in a protest movement against Prime Minister Benjamin Netanyahu.[28] The US Department of Homeland Security has relied heavily on commercial data brokers to maintain detailed location tracking on millions of immigrants, without obtaining the warrants they would need to collect such data directly.[29] Even digital snake oil is used toward antidemocratic ends; for instance, the European Union installed "smart lie-detection systems" at many of its busiest border crossings and has used the output of these pseudoscientific AI systems to target individuals for inspection and detention.[30]

Although human rights groups, journalists, and other advocates are pushing back against the erosion of democratic norms, the tide is currently against them. There are relatively few bulwarks preventing the secret life of data from being weaponized on a global scale, and every new technological innovation can feed into this process, accelerating the erosion. As we'll discuss below, hacking tools like Pegasus aren't the only challenge; increasingly, technology created to achieve socially beneficial ends is being repurposed to serve the digital cold war and to further undermine human rights and democratic norms around the world.

A HACK IN SHEEP'S CLOTHING

In light of the digital cold war and the many other societal challenges we've described in this book, it may be tempting to paint all information technology as inherently harmful, morally wrong, and antidemocratic. But that's only half the story; like so many other aspects of the human condition, tech is a double-edged sword that can also be wielded to strengthen democracies, advance social justice, and empower communities to envision and build a better future together. In fact, there is a large and growing worldwide civic tech movement that includes thousands of projects to improve social justice, human rights, and quality of life. Some such initiatives are public Wi-Fi access, government transparency tools, and open-source sensor networks tracking air and water quality on a global scale.

Yet for all the significant threats and benefits it offers to society at large, information technology has become so deeply integrated into our daily routines that its operations have become almost invisible. We have stopped thinking of it as a work in progress and started treating it almost like a natural resource, as abundant and necessary as air or water. Data processing is now an intrinsic dimension of our intimate lives, deeply enmeshed in our bodies, our bedrooms, and in the bonds between us. So, whether we think of tech as harmful or helpful, we're tempted to throw up our hands and accept its consequences as necessary and inevitable.

None of us has the power to turn the internet off and start over. And even if we could, let's be honest—most of us probably wouldn't flip the switch. It's often hard enough just to put our phones down and look our friends, family, and colleagues in the eyes. Technology's global reach, and our intimate dependence on it, can make us feel powerless even as we're complicit in perpetuating its central role in our society. So it's tempting to lay the blame for what ails our democracy at the feet of tech itself, because that lets *us* off the hook for doing something about it.

But it's both facile and disingenuous to think of these challenges as something happening independently of our own decisions, institutions, and politics. If we really want to fix the challenges that tech poses to democracy, we need to start paying more attention to the people who build and use it, to their overt agendas and implicit assumptions, and to potential alternatives. We need to shift the analytical frame from what went wrong to what we might do differently.

A great starting point for this shift in perspective is to look at cases where information technology that was designed, marketed, and adopted for one purpose is adapted to serve a completely different end—where people have decided to use tech to serve new agendas, often without the knowledge or consent of those affected. We've discussed this phenomenon briefly in other chapters, using the metaphor of the Trojan horse. Of course, technology has been repurposed since long before the digital era. The Slinky, one of the world's most popular children's toys, was originally designed in 1943 as an industrial spring for naval vessels; Silly Putty, another perennial playroom staple, was invented the same year as a synthetic rubber replacement to aid the US war effort in the face of a shortage due to Japanese domination of rubber-producing nations in the Pacific region. But internet-based products and services are uniquely prone to repurposing because they are networked technologies, and therefore their manufacturers—or any other parties with access to the servers on which their code and data reside—can exploit them, long after the technology has been marketed, distributed, and adopted, to suit very different ends than their users might expect.

Trojan horses may be used to serve socially beneficial purposes. For instance, after Russia invaded Ukraine in 2022, the deputy mayor of Kyiv, Petro Olenych, decided to repurpose the city's official smartphone app, Kyiv Digital, to help residents cope with the assault. Though the app was originally intended to address mundane aspects of city life like paying parking tickets, its updated version gave users information about the location of bomb shelters and basic staples like food and medicine, as well as real-time alerts about air raids and other emergency situations. This is hardly a feel-good story; while the histories of Slinkys and Silly Putty have a real swords-into-plowshares feeling, the transformation of Kyiv Digital is a move in the opposite direction. As Oleg Polovynko, the city council's IT director, told *Time* magazine in an article about the app, "Now we're in a new IT age, where we need to put all of our technology minds towards military goals."[31] But at least in this case, the militarization of civilian technology was oriented toward saving lives, with the full knowledge and consent of its users.

Sadly, when it comes to Trojan horses, Kyiv Digital is the exception rather than the rule. In most cases, commercial and especially state institutions are exploiting the malleability of code and data to make consumer technology suit their own agendas, without adequately warning end users who have already integrated these devices, apps, and services into their daily lives. We've described how the US Department of Homeland Security buys historical location data about American citizens and residents, without a warrant, from third-party data brokers who harvest these data from thousands of everyday smartphone apps—many of which collect detailed records of their users' locations for no apparent reason other than to resell them on the open market. The same technique (and, in some cases, the same data) is used by military organizations for strategic purposes. For instance, *Vice* magazine reported in 2021 on a tool called Locate X that aggregates commercial location data and allows members of the US armed forces to use those data for "intelligence, surveillance, and reconnaissance as well as dynamic execution of targets" by armed, flying drones.[32]

In cases where data brokers can't deliver the goods, military and intelligence organizations are using gig work apps to harvest these data on their own. You may have used such apps yourself. Millions of people rely every day on apps like Fiverr and TaskRabbit to match workers and employers for short-term projects, whether it's installing a toilet, designing a logo, or picking up the dry cleaning. But sometimes the tasks available via such apps can serve very different interests. As the *Wall Street Journal* reported in 2021, a specialized gig work app called Premise pays individual users in nations around the world as little as ten cents per task to complete short information-gathering missions in their hometowns, such as photographing strategically important sites, without ever informing these workers that their actual employer is the US government.[33]

Domestic government agencies also engage in Trojan horse tactics with troubling regularity. Do you have a General Motors car with OnStar, the company's subscription-based mobile services platform? You might be paying for features such as turn-by-turn navigation, roadside assistance, and remote key fob, but as it turns out, you're also providing data to US Immigration and Customs Enforcement (ICE) and other law enforcement organizations, so they can track your location in real time if they suspect you've broken a law.[34] And it's not only law enforcement; the US Centers for Disease Control and Prevention (CDC) used an app called SafeGraph, similar to Locate X, in order to track millions of Americans without telling them, in part to assess whether and where people were complying with COVID-19 curfews.[35]

Police in Germany collected data from a COVID-19 contact-tracing smartphone app called Luca in order to track down potential witnesses to a crime. Although this abuse of data privacy is a clear violation of German law, the app's developer told the *Washington Post* that they handed over the data because "the health department probably simulated an infection under pressure or requests from the police."[36] And, according to two separate 2022 reports by the Center for Democracy and Technology and *The Guardian*, countless schools across America that installed surveillance systems to look for guns pivoted to using them for COVID-19 compliance tracking—and vice versa.[37] To make matters worse, in many cases, systems installed in schools for either or both of these public health and safety reasons are now being used principally for student disciplinary action that has nothing to do with either.

Our point here isn't merely that governments and businesses are partnering to surveil people around the globe at an unprecedented scale—that's certainly the case, as you already know if you've read this far. Our point is that *every* source of information about individuals' identities, behaviors, and locations, regardless of its stated purpose or original function, may eventually become a tool of state surveillance, an instrument of institutional discipline, or a vulnerable vector for attack, thanks to the secret life of data. There's no such thing as an app, online service, or networked device that *can't* be turned into a Trojan horse after you've installed it. And the temptation for governments of even the strongest democracies to abuse their access to such data, even in violation of their own privacy laws, is too great for us to ignore. How can democratic norms, predicated on a stable balance of informational power, possibly withstand the lopsidedness of a world filled with Trojan horses and populated by millions of people who unwittingly invite them into our lives?

THE AGE OF UNREASON

One of the foundational concepts in modern democracies is what's usually referred to as the *marketplace of ideas*, a term coined by political philosopher John Stuart Mill in 1859, though its roots stretch back at least another two centuries.[38] The basic idea is simple: in a democratic society, everyone should share their ideas in the public sphere, and then, through reasoned debate, the people of a country may decide which ideas are best and how to put them into action, such as by passing new laws. This premise is a large part of the reason that constitutional democracies are built around freedom of speech and a free press—principles enshrined, for instance, in the first amendment to the US constitution.

Like so many other political ideals, the marketplace of ideas has been more challenging in practice than in theory. For one thing, there has never been a public sphere that was actually representative of its general populace. Enfranchisement for women and racial minorities in the United States took centuries to codify, and these citizens are still disproportionately excluded from participating in elections by a variety of political mechanisms. Media ownership and employment also skews disproportionately male and white, meaning that the voices of women and people of color are less likely to be heard. And, even for people who overcome the many obstacles to entering the public sphere, that doesn't guarantee equal participation; as a quick scroll through your social media feed may remind you, not all voices are valued equally.

Above and beyond the challenges of entrenched racism and sexism, the marketplace of ideas has another major problem: most political speech isn't exactly what you'd call reasoned debate. There's nothing new about this observation; twenty-four centuries ago, the Greek philosopher Aristotle argued that *logos* (reasoned argumentation) is only one element of political rhetoric, matched in importance by *ethos* (trustworthiness) and *pathos* (emotional resonance). But in the twenty-first century, thanks to the secret life of data, *pathos* has become datafied, and therefore weaponized, at a hitherto unimaginable scale. And this doesn't leave us much room for logos, spelling even more trouble for democracy.

An excellent—and alarming—example of the weaponization of emotional data is a relatively new technique called *neurotargeting*. You may have heard this term in connection with the firm Cambridge Analytica (CA), which briefly dominated headlines in 2018 after its role in the 2016 US presidential election and the UK's Brexit vote came to light.[39] To better understand neurotargeting and its ongoing threats to democracy, we spoke with one of the foremost experts on the subject: Emma Briant, a journalism professor at Monash University and a leading scholar of propaganda studies.

Neurotargeting, in its simplest form, is the strategic use of large datasets to craft and deliver a message intended to sideline the recipient's focus on logos and ethos and appeal directly to the pathos at their emotional core. Neurotargeting is prized by political campaigns, marketers, and others in the business of persuasion because they understand, from centuries of experience, that provoking strong emotional responses is one of the most reliable ways to get people to change their behavior. As Briant explained, modern neurotargeting techniques can be traced back to experiments undertaken by US intelligence agencies in the early years of the twenty-first century that used functional magnetic resonance imaging (fMRI) machines to examine the brains of subjects as they watched both terrorist propaganda and American counterpropaganda. One of the commercial contractors working on these government experiments was SCL, the parent company of CA.

A decade later, building on these insights, CA was the leader in a burgeoning field of political campaign consultancies that used neurotargeting to identify emotionally vulnerable voters in democracies around the globe and influence their political participation through specially crafted messaging. While the company was specifically aligned with right-wing political movements in the United States and the United Kingdom, it had a more mercenary approach elsewhere, selling its services to the highest bidder seeking to win an election.[40] Its efforts to help Trump win the 2016 US presidential election offer an illuminating glimpse into how this process worked.

As Briant has documented, one of the major sources of data used to help the Trump campaign came from a "personality test" fielded via Facebook by a Cambridge University professor working on behalf of CA, who ostensibly collected the responses for scholarly research purposes only. CA took advantage of Facebook's lax protections of consumer data and ended up harvesting information from not only the hundreds of thousands of people who opted into the survey, but also an additional 87 million of their connections on the platform, without the knowledge or consent of those affected. At the same time, CA partnered with a company called Gloo to build and market an app that purported to help churches maintain ongoing relationships with their congregants, including by offering online counseling services. According to Briant's research, this app was also exploited by CA to collect data about congregants' emotional states for "political campaigns for political purposes." In other words, the company relied heavily on unethical and deceptive Trojan horse tactics to collect much of its core data.

Once CA had compiled data related to the emotional states of countless millions of Americans, it subjected those data to analysis using a psychological model called OCEAN—an acronym in which the *N* stands for *neuroticism*.[41] As Briant explained, "If you want to target people with conspiracy theories, and you want to suppress the vote, to build apathy or potentially drive people to violence, then knowing whether they are neurotic or not may well be useful to you."

CA then used its data-sharing relationship with right-wing disinformation site Breitbart and developed partnerships with other media outlets in order to experiment with various fear-inducing political messages targeted at people with established neurotic personalities—all, as Briant detailed, to advance support for Trump. Toward this end, CA made use of a well-known marketing tool called A/B testing, a technique that compares the success rate of different pilot versions of a message to see which is more measurably persuasive.

Armed with these carefully tailored ads and a master list of neurotic voters in the United States, CA then set out to change voters' behaviors depending on their political beliefs—getting them to the polls, inviting them to live political events and protests, convincing them *not* to vote, or encouraging them to share similar messages with their networks. As Briant explained, not only did CA disseminate these inflammatory and misleading messages to the original survey participants on Facebook (and millions of "lookalike" Facebook users, based on data from the company's custom advertising platform), it also targeted these voters by "coordinating a campaign across media" including digital television and radio ads, and even by enlisting social media influencers to amplify the messaging calculated to instill fear in neurotic listeners. From the point of view of millions of targeted voters, their entire media spheres would have been inundated with overlapping and seemingly well-corroborated disinformation confirming their worst paranoid suspicions about evil plots that only a Trump victory could eradicate.

Although CA officially shut its doors in 2018 following the public scandals about its unethical use of Facebook data, parent company SCL and neurotargeting are still thriving. As Briant told us, "Cambridge Analytica isn't gone; it's just fractured, and [broken into] new companies. And, you know, people continue. What happens is, just because these people have been exposed, it then becomes harder to see what they're doing." If anything, she told us, former CA employees and other, similar companies have expanded their operations in the years since 2018, to the point where "our entire information world" has become "the battlefield."

Unfortunately, Briant told us, regulators and democracy watchdogs don't seem to have learned their lesson from the CA scandal. "All the focus is about the Russians who are going to 'get us,'" she said, referring to one of the principal state sponsors of pro-Trump disinformation, but "nobody's really looking at these firms and the experiments that they're doing, and how that then interacts with the platforms" with which we share our personal data daily.

Unless someone does start keeping track and cracking down, Briant warned, the CA scandal will come to seem like merely the precursor to a wave of data abuse that threatens to destroy the foundations of democratic society. In particular, she sees a dangerous trend of both information warfare and military action being delegated to unaccountable, black-box algorithms, and "you no longer have human control in the process of war." Just as there is currently no equivalent to the Geneva Conventions for the use of AI in international conflict, it will be challenging to hold algorithms accountable for their actions via international tribunals like the International Court of Justice or the International Criminal Court in The Hague.

Even researching and reporting on algorithm-driven campaigns and conflicts—a vital function of scholarship and journalism—will become nearly impossible, according to Briant. "How do you report on a campaign that you cannot see, that nobody has controlled, and nobody's making the decisions about, and you don't have access to any of the platforms?" she asked. "What's going to accompany that is a closing down of transparency . . . I think we're at real risk of losing democracy itself as a result of this shift."

A PERFECT STORM IS BREWING

Briant's warning about the future of algorithmically automated warfare (both conventional and informational) is chilling and well founded. Yet this is only one of many ways in which the secret life of data may further erode democratic norms and institutions. We can never be sure what the future holds, especially given the high degree of uncertainty associated with planetary crises like climate change. But there is compelling reason to believe that, in the near future, the acceleration of digital surveillance; the geometrically growing influence of AI, ML, and predictive algorithms; the lack of strong national and international regulation of data industries; and the significant political, military, and commercial competitive advantages associated with maximal exploitation of data will add up to a perfect storm that shakes democratic society to its foundations.

One rapidly emerging threat to democratic norms is all too familiar from science fiction, most prominently *Minority Report*, the blockbuster 2002 film based on a 1956 short story by the visionary author Philip K. Dick. In the film, a group of psychic "precogs" employed by the police department's "precrime" division predict the nature, location, and perpetrators of specific crimes, allowing the police to apprehend future criminals before they commit their heinous acts. The story has become something of a touchstone for critics of technology and society, because several plot points that seemed like pure dystopian fantasy in 2002 have emerged as viable or marketable technologies and practices a mere two decades later. Chief among these is the concept of precrime enforcement, which is practiced in the United States and elsewhere as *predictive policing*.

The idea that people are innocent until proven guilty dates back at least three centuries, to when an anonymous London barrister with the initials L. M. wrote in a 1725 political pamphlet that "every Person ought to be supposed innocent, until the Contrary appears."[42] It's a foundational premise of modern democracy, and a deeply shared common value worldwide.[43] Yet predictive policing, which uses algorithmic assessments to justify enforcement actions against people who are expected to commit a crime, rather than suspected of having committed one, has grown increasingly widespread in the United States and elsewhere over the past decade. Now, millions of people who haven't done anything illegal are effectively treated as guilty until they're proven innocent.

Software companies like Vaak in Japan and Third Eye in the United Kingdom have marketed services to retail businesses that purport to spot shoplifters before they steal, using predictive analytic models based on surveillance data. American companies like ShadowDragon, Palantir, Kaseware, and PredPol sell software and services to countless law enforcement agencies, promising to do the same thing on a municipal scale—identifying targets for surveillance and detention before any crimes have been committed, based on proprietary algorithms that combine multiple sources of data with unregulated and unaudited projective modeling programs.[44] Some local law enforcement agencies, such as the Pasco county sheriff's office in Florida and the Chicago police department, have built their own such systems in-house. And as a 2022 investigation by the *New York Times* revealed, the Chinese government has been quietly building its own predictive policing system on a national scale.[45] Its approach to proactive control over criminal intent is so all-encompassing that several novelists in the country have reportedly found their work censored and deleted by cloud-based word processing programs based on suspicions that they were planning to publish information deemed illegal under Chinese law.

Based on what you've read so far, you may be wondering: Do the issues of algorithmic bias that plague so many ML and AI systems play a role in predictive policing? The answer, sadly, is *of course they do*. Predictive policing algorithms are almost definitionally biased in ways that reflect and widen existing social inequities. A 2021 investigative report by Gizmodo analyzed over seven million crime predictions from PredPol, spanning thirty-eight American cities and counties over the course of the previous three years. Among other things, the report found that "PredPol's algorithm relentlessly targeted [neighborhoods that were] the most heavily populated by people of color and the poor" in a way that was disproportionate to actual patterns of crime. To make matters worse, Gizmodo found, "arrests of and police use of force on people of color were much more prevalent in the areas that PredPol targeted most frequently" than in similar neighborhoods, strongly suggesting that the use of the software exacerbated legacy systems of biased policing, making living conditions worse for people who were already the most vulnerable to systemic abuse.[46]

We interviewed Paul Hetznecker, a civil rights and criminal defense attorney who has litigated many predictive policing cases during his three decades of private practice. He believes these biased outcomes are not accidental, regrettable oversights in an otherwise benevolent policing system, but are rather the result of new data technologies bolstering long-standing systems of oppression. In his words, predictive policing is "just a high-tech way of profiling" minorities and the poor. What's new, he told us, is the ability to use computer modeling to paint a fig leaf of objectivity over the process. This fig leaf, in turn, has enabled a shift in policing strategy from reactive to proactive—a model of social control that was widely seen as unconstitutional before the existence of this technological rationale.

Yet in the final analysis, Hetznecker told us, the inputs into these supposedly objective systems are exactly the same as the factors that would have led a police officer to profile someone for surveillance, harassment, or detention a century ago: "The way police approach 'future crime' is they look at behavior. They look at association, at geography, which incorporates class and race. All of these factors utilized by police through these AI predictive policing programs continue the traditional model of repressive and racist 'warrior policing' policies."

The logic of predictive policing isn't limited to law enforcement, and the consequences of algorithmic bias in predictive modeling extend far beyond the criminal justice system. If anything, Philip K. Dick's dystopian vision of a world in which the punishment precedes the crime falls short of our current reality (though some episodes of the television show *Black Mirror* come closer). In the 2020s, billions of people living in nations all over the planet are not only continuously surveilled and profiled, but are also scored, ranked, rewarded, and punished based on surveillance data and analysis, often without knowing how or why.

To date, the most ambitious example of this process is China's social credit system (SCS), which was first announced by the state council in 2014 and has been in development since then. Though it's shockingly difficult to get details about the extent of this system, most reliable analyses agree on at least a few key points.[47] The SCS uses data inputs from business transactions, local governments, internet services, financial services, video surveillance networks, facial recognition systems, and other sources, then centralizes these disparate elements in a database controlled by the national government. Local governments, law enforcement agencies, and other authorities are granted the ability to query this database, and selective access to the data it contains.

Though there may or may not be a single, dynamic score assigned to each Chinese national based on the algorithms applied to the SCS database (reports differ on this point), it is used to judge the overall "trustworthiness" of citizens. Those deemed highly trustworthy may be placed on a "red list" and granted certain social privileges including financial incentives, while those deemed to have a low level of trustworthiness are placed on a "blacklist" and may be punished. To date, punishments for blacklisted individuals and their families have reportedly included public humiliation on video billboards, restrictions on travel, exclusion from buying luxury goods and services, exclusion from employment and educational opportunities, slower internet speeds, and enhanced surveillance, including regular in-person visits from law enforcement.[48]

Much of the coverage of the SCS in Western media has an element of self-congratulatory horror and relief ("How barbaric! At least it's not happening here," seems to be the subtext). Yet, as Curtin University researchers Karen Li Xan Wong and Amy Shields Dobson argued in a 2019 article in the academic journal *Global Media and China*, "the infrastructural and cultural foundations for a social credit system exist in Western democratic countries," as well.[49] Those foundations may not be centralized by a national government, but there's so much cross-pollination between commercial and government data, and so much coordination between systems of financial, cultural, and state power, that the consequences can be much the same. Between credit scores, Yelp ratings, social media follower counts, quantified self records, LinkedIn pages, data broker profiles, and countless other publicly available and privately assembled dockets and dossiers, residents of the United States, European Union, and other liberal democracies may be just as likely to find themselves rewarded and punished by unaccountable algorithms crunching the numbers as Chinese citizens living under the SCS regime. And the rewards bestowed by this system are far more likely to fall on the plates of those who already benefit from a higher social status, while the punishments will continue to impact women, minorities, and low-income individuals, families, and communities to a disproportionate degree.

GLIMMERS OF HOPE

How can democracy survive the secret life of data? Between the digital cold war, rampant Trojan horse apps, neurotargeting, predictive policing, and social credit scores, it's hard to imagine a world in which civil liberties, government accountability, and social equality can flourish. And, we're sorry to say, the cases and scenarios in this chapter barely begin to capture the full scope of these threats and challenges, which—like information technologies—continue to evolve a lot more quickly than researchers can write books about them.

And yet there's good reason not to abandon all hope for democracy. It's a resilient political and social system, in part because it's built on the assumption that it will always be incomplete, imperfect, and vulnerable to hacking. That gives those of us who live in democratic societies both the confidence and the tools to fix problems when we see them, rather than waiting helplessly while the system falls to pieces around us.

There's a lot we can do, should do, and are doing, right now, to help prevent the worst from happening. Because these threats come at the intersection of technology, politics, and industrial capitalism, we must work on all three fronts simultaneously.

One of the most important things democratic societies can do is to regulate data industries and the use of data as a commodity. This is easy to say, but it's no easy task; some of the leading lights in international scholarship and policy have published entire books about single clauses of the technology laws of their countries, and there's no way to approach the subject with both nuance and comprehensiveness (which opens the door to a lot of useless bloviation, false promises, and quick "fixes" that may only make matters worse).

Yet, as a result of sustained research and advocacy, including by some of the people whose voices you've heard in this book, things are beginning to change. In the United States, as we were writing this chapter, two members of Congress who chair powerful committees wrote a letter to the US attorney general and the heads of several federal agencies, including the Department of Justice, the Federal Bureau of Investigation, and the Department of Homeland Security, requesting that these agencies account for their dealings with data brokers. The letter expressed concern that "improper government acquisition of this data can thwart statutory and constitutional protections designed to protect Americans' due process rights."[50] While this is hardly binding policy, it's a sign that lawmakers are aware of the problem and willing to flex their muscles to fix it.

More concretely, the Federal Trade Commission (FTC) has also responded to pressure from public advocates to rein in the unchecked power of profit-hungry tech firms by regulating improper commercial uses of consumer data. A new policy framework is emerging that treats the provenance of data—how it was harvested, from whom, and under what circumstances—as an element of fairness in trade practices. To name one example, the FTC has issued numerous orders in recent years to tech firms—including Cambridge Analytica—requiring them to retire algorithms that were trained on illegally harvested data.

Other countries have also made strides toward limiting the impacts of data and algorithms on democratic systems and human rights. In the European Union, which passed landmark data privacy regulations in 2018 (partially in response to the Cambridge Analytica scandal), more recent efforts have aimed to curb the use of artificial intelligence. The AI Act, a sweeping piece of legislation proposed in 2021 and not yet passed into law at the time of writing, would ban certain uses of AI, including those with "potential to manipulate persons through subliminal techniques," those that "exploit vulnerabilities" of specific groups defined by age or ability, and those that evaluate the "trustworthiness" of people in order to develop a "social score" like China's SCS. Similar legislation is currently working its way through the Brazilian legislature as well, though it's unclear at this point whether the final version would have the same kind of enforcement power as the proposed EU regulation. Even China has cracked down on unethical commercial uses of algorithms, with new and proposed regulations that outlaw practices such as discriminatory pricing, algorithmically generated fake news, and deepfakes.

We see other glimmers of hope for democracy, not in the political process, but in the behavioral and cultural changes of people who live in data-saturated societies. There are countless promising signs that everyday people are becoming more aware of the secret life of data and developing techniques to limit its impact on their lives and communities.

Some of these changes have to do with how people use technology. For instance, the use of encryption, which allows internet users to transact data secure in the knowledge that third parties can't easily intercept it, has climbed significantly in recent years. Market research suggests that more than a billion internet users currently encrypt their traffic with a virtual private network (VPN) service, a number that has nearly doubled over the past five years.[51] Encrypted chat applications have also grown dramatically. Telegram had about 550 million monthly users by the middle of 2021—nearly double its user base in 2019, and more than ten times the number in 2014. Signal had 40 million users in January 2022, roughly twice as many as a year earlier. Figures like these are especially impressive, and inspiring, in light of efforts to restrict consumer uses of encryption in many nations around the world, under both authoritarian and democratic political systems.

Other shifts have less to do with the technology itself and more to do with how we interpret its role in culture. As we discussed in chapter 6, people who become used to seeing themselves refracted through the lens of data processing systems start to develop algo-vision—a knack for anticipating how their actions will be interpreted by the algorithms they interact with. This superpower is also being used to blunt the role of data in warping democratic society. On social media platforms like Instagram and TikTok, legions of media literacy advocates like commercial photographer Caroline Ross post educational videos demonstrating how celebrities and others manipulate media and teaching their millions of viewers how to spot deepfakes and Photoshopped images.

Along similar lines, there are millions of internet users who participate in open-source intelligence (OSint) efforts to hold governments and corporations accountable, using techniques analogous to the ones used by powerful institutions to keep tabs on citizens and consumers. For example, months before Russian president Vladimir Putin formally announced his invasion of Ukraine in 2022, a loose network of political scholars and citizen sleuths documented the buildup to war by monitoring traffic conditions on Google Maps and combing through hundreds of TikTok videos from users in Ukraine, Belarus, and Russia. Although the Russian military forbade its soldiers from using smartphones in an effort to curb leaks, it lacked the power to erase their digital footprints altogether. As Middlebury College professor Jeffrey Lewis explained to the *Daily Beast*, "We live in an era where all of our patterns of life are digitally recorded and recognized. Even if Russian troops turned their phones off, the disruption that they caused was a deviation from the normal pattern of life."[52]

We're not suggesting that any of these new laws, regulations, technologies, cultures, or behaviors will fix the problems facing democracy in the networked age. These are deep, systemic challenges, and if there were an easy answer, we wouldn't have needed to write this book, and you wouldn't need to read it. But ultimately, both technology and democracy are human systems, made by human beings to suit human purposes and to mediate between the conflicting demands of different human values and agendas. It's important to recognize that they exist because we exist, and that our everyday speech and behaviors help to shape and constrain their development and use, just as much as they shape and constrain our lives and societies.

We all have important roles to play in this reciprocal and continuing process. The more we inform ourselves and one another about where technology comes from and how it works, and the more we peel back the layers keeping the secret life of data secret, revealing the hidden circuitry of power so that we might intervene on behalf of human dignity and justice, the closer we'll come to realizing technology's promise and to building a world in which everyone has an opportunity to grow and thrive together.

Conclusion: Data Afterlives

This book began with a simple premise: that no matter how precise, robust, comprehensive, accurate, or self-evident a dataset may appear to be, it will always have a secret life. There is no algorithm, analytical model, or artificial intelligence that can predict or delimit the full extent of knowledge that humans (or other sentient beings) may generate from data, and the possibilities for new discoveries and paradigms grow exponentially with time and distance from the original data source. Data, fundamentally, have no fixed meaning.

This is not a mere intellectual novelty or a dollar store distillation of pithier observations made by mathematicians and scientists like Bertrand Russell, Kurt Gödel, and Werner Heisenberg. We didn't write this book to coin a term, score a point, or attract academic citations. We wrote this book to raise an alarm, because the implications of our premise fascinate and terrify us in equal measure.

Over the course of our brief lifetimes, humanity has pursued an unprecedented number of new data sources, tools for data processing, and social uses of data-driven insights. As the authors of this book came of age, so did the internet, and the years of our adult lives have thus far been measured out in iPhone upgrades, social media scandals, and cybersecurity breaches. As researchers and artists working at the nexus of technology and culture, we are witnesses to a profound data-driven transformation in the social fabric and lived experience of every member of industrialized society. We feel compelled to add our voices to the rising chorus of those asking whether, why, and how humanity should continue along this path.

Over the course of writing this book, as we sifted through thousands of pages of journalistic, scholarly, policy, technical, and science-fictional sources, and interviewed dozens of experts in fields from computer science to law to design, something we suspected intuitively at the outset of this project has been abundantly confirmed. Namely, *nobody* (including us) has a comprehensive understanding of how many types of data are currently being collected, in what volume, or by whom, or how all these data are being stored, analyzed, and used in decision-making processes. Nor can anyone predict what future uses they may be put toward, or what the longer-term social implications of these data systems and practices might be.

It's not for lack of interest or lack of effort; all of the people we interviewed are intelligent, learned, ethical, engaged, and curious about technology and society. Most of them have depths of expertise in their respective fields that we couldn't hope to fathom without years of study. Some of them have even considered the broader questions we posed about the larger-scale and longer-term consequences of their work and offered us new avenues of insight into these problems (though a surprising number of interviewees said they hadn't considered such questions, being beyond the range of their own professional and intellectual focus). But very few, if any, experts in fields such as computer science, information science, social science, or technology policy appear to be grappling in a comprehensive way with the full scope of social transformation we face as a result of the ubiquitous data collection and processing regimes we have described in this book.

As for the popular discussion of technology and society, accessible to laypeople via journalism, political debate, and marketing materials, the narrative is so piecemeal that the bigger picture is barely addressed at all. Each new legislative bill, zero day hack, AI-assisted media production tool, dystopian surveillance scenario, and biometric smart device is discussed on its own terms, as a standalone marvel or threat, without any connective tissue to attach it to the broader context that unites these phenomena; and each is forgotten once the news cycle refreshes, rarely to be mentioned again. For the most part, even longer-form books and documentaries addressing these issues for popular audiences tend to place technology along a spectrum from dystopian to utopian, which at its extreme ends promises either a bold new future in which technology magically allows humankind to transcend suffering and privation, or a surveilled hellscape in which artificial intelligence becomes a taskmaster for enslaved human masses. Too few of these narratives enable the reader or viewer to understand how data collection and analysis may actually affect their own lives over the long term, let alone inspire a sense of agency to do anything about it.

The problem, we believe, is one of scale. The breadth of technologies, conceptual tools, cultural specificities, and social institutions entangled with the secret life of data is so great that nobody could possibly possess the expertise to comprehend it in toto. That's why many experts tend to focus on areas in which they can understand enough to make a difference—whether it's the design of biomedical implants, the policies governing facial recognition technology, or the optimal use of mesh networking to track ecological changes in a forest environment. And yet, even if the scope of the challenge is too great for any of us to appreciate in its entirety, it nonetheless affects us all, in countless ways that manifest at the scale of individual lives, families, and communities. Without a framework of understanding to provide context for these obstacles, opportunities, and transformations when they appear before us, we are all doomed to play catch-up in perpetuity. That discourages us from thinking or behaving proactively, let alone making decisions about the role that data will play in our private and public lives.

We hope that, in some small way, our book can help address this problem. Our aim was to create some conceptual scaffolding to help readers put the relentless torrent of new information technologies, social practices, and cultural trends into perspective. We wanted to take the data points provided in news stories, press releases, and social media posts and weave them into actionable knowledge that feels relevant to your lived experience.

We knew at the outset that this would be a quixotic task: to include enough specificity and scholarly rigor to avoid adding to the flood of bullshit that sullies the discourse about data and society, while also writing something accessible and (we hope) fun to read for a general audience. Finally, we aim to raise some dire concerns about the future without seeming like Chicken Little or battering our readers into a depressive stupor. As we write this, we have no idea whether or not we've succeeded, but if we have, it's only because of the generosity and thoughtful contributions of the twenty-nine experts we interviewed, as well as our many friends and colleagues around the world who reviewed drafts in progress. Their diverse perspectives and deep wells of experience allowed us to triangulate a far larger picture of the world than we could have mustered from our own limited vantages.

We can't possibly claim to understand the full extent of the secret life of data, let alone the full scope of its social implications. Nobody could—that's kind of the point. But in the remainder of this conclusion, we'll do our best to wrap up a few of the major threads in this book, hopefully providing a starting point for our readers to exercise more agency and discretion in their own lives, while leaving some other threads hanging so that later scholars, artists, and authors can pick them up and weave them into something even better.

DATA, KNOWLEDGE, AND CULTURE

One of the main themes that emerged from our research for this book is that data are, by definition, cultural artifacts. Every aspect of data collection, analysis, and processing, as well as the journey from unstructured data to structured data to information and ultimately to knowledge, is circumscribed by cultural forces and factors, informed by cultural values, distorted by cultural biases, and finally fed back into the cultural process of data collection in a never-ending cycle. There is no such thing as data without culture—or vice versa. This helps to explain why the radical changes in our technological approach to collecting and analyzing data, spurred by rapid advances in computer processing and information networking, have such a profound capacity to alter our cultural landscape.

Nobody within arm's reach of an internet-connected laptop, smartphone, television, or IoT device could deny the dominant role that what used to be called "digital culture" plays in mainstream, mediated discourse in nearly every corner of the world. Memes can help turn the tide in national elections; social media influencers can make or break the fortunes of Fortune 500 companies; and, in many places, one well- or poorly planned tweet, post, or TikTok video can propel an individual to global celebrity or land them in prison.

AI-assisted production tools are, at the time of writing, all the rage, and GANs, deepfakes, and other media generated with the assistance of ML algorithms have flooded visual (and, to a lesser extent, musical) culture. The inevitable debates have followed: Who, if anyone, should own a piece of media generated by a computer algorithm with minimal input by a human? Should human artists have the power to restrict their work from ML training sets, so their stylistic individuality can't be imitated by machines—and, more importantly, so they can demand an economic stake in any algorithmically generated products that draw on their creative labor? Who should be held accountable for AI-generated disinformation and hate speech? These are all excellent questions, and the ways in which different societies choose to answer them will have consequences not only for art and artists, but for the emergent political, social, and economic structures that tend to follow major cultural shifts.

Yet, while the memeification and GANification of popular culture is perhaps the most visible cultural effect of the secret life of data, there are also more subtle, tectonic, and comprehensive ways in which new information technology has altered cultural processes. One is what you might call the *quantization of culture*. This is related to a theme we discussed at length in chapter 6: the human tendency to absorb the ways in which data systems "understand" them into the ways they understand themselves, leading to a kind of fractured consciousness that we call algo-vision.

One consequence of algo-vision is that we begin to define our identities and process our experiences through the categorical gradations that were created by third parties for the purposes of surveilling, predicting, and controlling our behavior. Human beings have done this since long before the internet, of course. Think about people self-identifying in generational cohorts ("I'm a Gen Xer, and millennials will never understand what makes classic hip-hop so great"), and consider how often you've heard a tech fan describe themselves as an "early adopter" to justify spending an exorbitant sum of money on a new device. These categorization systems, which were invented so that marketers could segment their audiences more effectively, don't actually describe or predict people's identities and behaviors in any meaningful way—except in cases where we internalize their logic and then change ourselves to conform to their dictates.

In the 2020s, because so many of our social interactions and cultural behaviors are now mediated through algorithmic processes, this happens far more frequently, and less noticeably, than it used to. At the time of writing, for instance, Facebook allows users to react to other posts with a choice of seven different emotional categories. And, as the *Washington Post* reported in 2021, the service's algorithm doesn't treat all of these responses with equal weight; leaked internal documents showed that a post with an "angry" reaction was five times more likely to be promoted in users' feeds than one with a "like" reaction.[1] In other words, this algorithm doesn't merely require users to reduce their complex emotional palettes to a sparse set of seven expressive options, but it also rewards us more for inspiring anger than for likeability.

Whether they are aware of it or not, users tend to internalize the logic of these systems as they interact with them, and they begin to experience and express emotions through these categories and quantities, tailoring their self-presentation to optimize the chances of scoring points and gaining visibility. Ironically, when this happens, it has the unintended consequence of validating the categorization scheme adopted by the algorithms' creators ("See? Millions of people find the hate button useful!"), a process analogous to what statisticians call *confirmation bias*. This can be understood as yet another case of a feedback loop between cultural processes and algorithmic ones.

Another cultural consequence of the secret life of data is even more subtle, but potentially transformative, and we refer to this as the *erosion of certainty*.[2] This is also the continuation of a longstanding trend; in the twentieth century, as microscopes and mathematics reached new levels of granularity, scientists and philosophers began to understand that there are fundamental limits to what we can know (exemplified by Heisenberg's famous uncertainty principle), and that these limits cannot be overcome simply by continuing to refine our measurement tools. Similarly, as our digital tools for data collection and analysis continue to improve in the twenty-first century, these new digital techniques paradoxically make it more difficult to arrive at definitive conclusions and build reliable knowledge.

One example is something that's widely referred to in scholarly circles as the *reproducibility crisis*. Reproducibility is what makes the scientific method scientific: If I measure the weight of a particle or the size of a star, using the same methods and data that you used, I should come up with the same results that you did. But new, data-centric modes of knowledge production are challenging reproducibility in at least two ways.

First, as it turns out, a lot of published scientific studies that authors claimed were reproducible can't actually be reproduced. Over the past decade or so, researchers have used new analytical techniques to reexamine old datasets, and also to test the internal logic of the figures and statistics included in scientific papers. In many cases (possibly most cases), it turns out that the original numbers were fudged, poorly measured, misrepresented, or misunderstood. This means that all of the subsequent studies done by other scholars based on these flawed findings were also incorrect, making them as shaky as the upper floors of a skyscraper built on quicksand. To make matters even worse, researchers from University of California San Diego found in a 2021 meta-analysis that nonreplicable studies are cited *more often* than reproducible ones, compounding the harm even further.[3]

You might be thinking it's a good thing that all of these flawed findings are coming to light, thanks to the closer scrutiny provided by new tools for data analysis. Sure, it forces us to throw out a lot of what we thought we knew, but once we get rid of all the junk, what's left standing will be even stronger, right? Well, not quite. That would be true if we weren't also using the same kind of big data tools to generate a larger and larger portion of new scientific knowledge. As it turns out, new machine learning and artificial intelligence techniques are also prone to fudging, errors, and misinterpretation, but they're much harder to hold accountable than the older techniques, because they're often reliant on black-box algorithms that few (if any) of their users fully understand, and neither their inner workings nor their training datasets can be easily audited (let alone reproduced) by third parties. This is yet another self-compounding problem: as Gabriel Grill, a doctoral candidate at the University of Michigan, has shown, computer scientists and researchers who rely on algorithmic analysis often overstate their confidence in these computational models in order to ward off criticism, presenting AI and ML as unassailably accurate ways of understanding the world when in fact they're often no better at producing reliable knowledge than older, more transparent and auditable statistical tools.[4]

The quantization of culture isn't just something that happens on social media, and the erosion of certainty isn't just a problem for scientific research and AI models. As our research for this book has shown us, these two contradictory trends are even more powerful in combination than on their own. The more we integrate computational modeling into our understanding of our lives and the world around us, allowing algorithms to intervene in our interpersonal relationships and innermost experiences, the less confidence we have in the authenticity of our identities, emotions, and environments, and in our ability to make decisions for ourselves. This lingering sense of uncertainty and indeterminacy, even as everything else about our lives, from our heart rates to our geographical locations to our credit scores, is quantified, cataloged, and updated by the millisecond, can leave us with a profound sense of paralysis and helplessness that makes it difficult to maintain a conversation, let alone a democratic society.

SLOW FIXES

We would love to conclude this book with a quick fix for all the challenges we've raised in the preceding pages—a handy one-size-fits-all solution for any problem that might crop up thanks to the secret life of data. We're sorry to tell you that we don't have one, and neither does anybody else (regardless of what they might say in their TED talk, business pitch, or op-ed). But that doesn't mean we should all just give in to despair and allow the world around us to devolve into a bad *Black Mirror* episode (or, worse, a good one). There are a lot of smart, ethical people working hard to chart a path that will help information technology benefit the human condition while minimizing some of its most troubling consequences. And while there might not be a single quick fix, there are myriad potential slow fixes, each of which can play a role in promoting human agency, justice, and equity in a data-rich society. Many of these approaches fall into three categories: new laws and regulations, new ethics and aesthetics, and new models of production.

As we discussed in chapter 8, there are a host of new laws around the world aiming to limit the commercial scope of data collection and analysis and to crack down on the use of AI and ML algorithms to make decisions that affect people's social, political, and economic circumstances. At the time of writing, for instance, the US House of Representatives is considering a bill called the American Data Privacy and Protection Act (ADPPA), which would limit commercial data collection to what is "reasonably necessary" in order to deliver digital services and give people the ability to audit and delete their data, as well as the choice to opt out of targeted advertising. Like the European Union's General Data Protection Regulation (GDPR), a similar policy that went into effect in 2018, this law would give state and federal regulators the power to enforce its provisions with fines, as well as permit individuals whose data privacy has been violated to bring civil actions against companies that have abused their data. On the other hand, much to the consternation of privacy and human rights activists, it would lack some additional key consumer protections, just as the GDPR does.

Laws and regulations like the ADPPA and GDPR focus on commercial entities' collection and storage of consumer data, using economic sanctions to make it costlier for companies to indiscriminately harvest and retain every piece of data available (a widespread practice described to us by Foursquare founder Dennis Crowley, as discussed in chapter 4). But other new and pending regulations focus on what comes next in the process: the AI and ML algorithms themselves, which use bulk data to generate new models and guide important decision-making processes. Thus far, most successfully implemented laws and policies—from the nation of Morocco to the US state of Colorado—tend to focus narrowly on a single AI-related issue, such as reining in the uses of facial recognition technology. Broader bills like the Algorithmic Accountability Act in the United States and the AI Act in the European Union aim to place more sweeping limits on the uses of algorithms in public life and commerce, and their more ambitious scope might help to explain why, at the time of writing, no such legislation has yet passed.

Nearly all of these proposed and implemented bills, laws, regulations, and policies aiming to blunt the social impact of data and algorithms are limited in one way or another, whether by scope (focusing on a single issue), by purview (carving out exemptions for law enforcement surveillance or "necessary" commercial uses), or by enforceability (restricting the ability of individuals to bring civil actions against data abusers). All of them, as far as we can tell, are *reactive* in nature, responding to specific perceived threats with policy interventions meant to limit their harm. Yet, for years, there have also been calls for *proactive* policies to establish affirmative human rights related to data, applicable across a wide spectrum of potential abuses.

Among the many voices clamoring for such affirmative rights is the White House Office of Science and Technology Policy in the United States, which in 2022 released a statement titled "Blueprint for an AI Bill of Rights."[5] Shortly thereafter, in 2023, President Biden issued an executive order instructing federal agencies to marshal civil rights authorities in order to crack down on algorithmic discrimination and other potential social harms associated with data and AI.[6]

While it's heartening to see policymakers acknowledging the importance of affirmative rights related to data, it remains to be seen whether and how these proclamations will translate into actionable policy. This is due in part to the fact that executive decisions have less persuasive force and less staying power than binding legislation in the American political system. It is also due in part to the fact that AI's attendant risks and rewards, and the optimal balance between the two, appear very differently to different stakeholders (as we witnessed in our interviews throughout the writing of this book). Yet another facet of the challenge is the significant degree of self-interest in both the commercial and governmental sectors; the power to surveil, predict, and influence billions of people using algorithms is so valuable that it's difficult to give up voluntarily. But part of the challenge is also conceptual: if we can't predict the secret life of data, how can we avoid its many potential pitfalls through proactive policy?

Many erudite legal scholars have offered compelling answers to this question, adapting centuries of human rights law, property law, trade law, and communications law to encompass new forms of data-driven power. But like all fundamentally transformative approaches to governance, data rights occasion more than a simple question of figuring out the legal technicalities and specificities. Before we can agree on a policy framework, we need to agree on some basic principles. We need to know what we're trying to achieve, at the greatest scale, over the longest term. In other words, we need an *ethical* framework.

Fortunately, there are many to choose from. Data ethics has become something of a boom industry for social theorists and philosophers in recent years, buoyed in roughly equal measures by legitimate concern about the dangers of unchecked surveillance and algorithms and a flood of money from private-sector organizations hoping to demonstrate their commitment to ethics without having to make economic sacrifices or face binding regulation. While we can't provide a categorical overview of every such ethical framework, and we won't pick favorites, there are some useful approaches that stand out in light of the issues raised during our research, which we'll describe briefly below.

One framework that has emerged over the past decade is what's often referred to as *data justice*—a term that was coined by cybersecurity consultant Eleanor Saitta on Twitter in 2012 and absorbed rapidly into scholarly and policy circles.[7] Though various scholars and activists have since defined it differently, most focus on the question of "how social justice can be advanced in a datafied society," as the founders of the Data Justice Lab at Cardiff University have written.[8]

Data justice has two complementary dimensions. From one perspective, this ethical framework allows people to critique the ways in which data systems affect societal justice, for instance by exacerbating racial divides through algorithmic bias, as we discussed in chapter 3. But from another perspective, the concept also allows activists to bring data systems to justice by identifying the stakeholders who build them, deploy them, and use them to make decisions that have social impacts. Both of these perspectives are useful in crafting policy to address some of the potential harms of the secret life of data, because they demystify technology and treat it as a social institution instead of a world unto itself. However obscure and seemingly autonomous an algorithm might be, tucked inside a black box of proprietary code or protected as a powerful company's intellectual property, it can still be understood tactically as a way for some people to influence the lives of others—and, therefore, people can be held accountable for its consequences.

A separate but related ethical value is what is often referred to as *data sovereignty*. This framework, which gained currency in the post-Snowden era, focuses on the intersection between data systems and political autonomy. As the NSA materials leaked by Snowden demonstrated, the centrality of US-based networks and services to internet usage around the globe amounts to a vulnerability for every other nation by undermining the ability of sovereign states and their citizens to shield their data from political surveillance and commercial exploitation. Many of the biggest changes to the structure and governance of the internet since then—from the construction of new regional networks and exchanges to the passage of new laws and regulations like the GDPR and Brazil's Marco Civil da Internet—are rooted in the ethic of data sovereignty and designed specifically to prevent the United States and other powerful nations from usurping political power through the collection and analysis of local data.

Like many other powerful political and ethical concepts, data sovereignty has continued to evolve beyond the circumstances from which it first arose. There is a loose worldwide coalition in support of indigenous data sovereignty, reflecting the distinct political challenges and assertion of rights associated with indigenous peoples in every populated region of the globe.[9] Though their circumstances and values differ widely, indigenous activists from Australia to North America to the Amazon tend to agree that data surveillance and algorithms can serve as instruments of power, reinforcing systems of dominance, subjugation, and extractive economic exploitation that date back centuries. Only through acknowledging these inherent threats to indigenous sovereignty, activists argue, can we develop laws and technologies that protect the rights of these communities from further injury. And, as with other issues that arise in conversations about indigenous rights, these critiques may have larger implications for other data sovereignty movements around the world, as they highlight important concepts about power and its unequal distribution in data's secret lives.

Another ethical framework that has gained traction in recent years is typically referred to as *participatory design*. As design professor and disability rights advocate Laura Forlano (who discussed her poorly designed medical implant in chapter 6) explained to us, the crux of the concept is in building "more equitable relationships" between technology designers and users. While the ethic of participatory design predates the commercial internet by decades, it has gained new currency and developed new urgency in the age of ubiquitous data networks and implantable biometric devices. As Forlano's latest scholarly work argues, it's not enough to simply include users in the process of designing the technologies that will shape their lives (though even that is still far too rare!). Rather, she writes, technology design must "center systematically excluded individuals and groups to address social injustice."[10] In other words, technology makers should focus on how their work will impact their most vulnerable users and include them in the design process on an ongoing basis. If they don't, the tech is virtually guaranteed to inflict physical and social harm on the people who most need it to benefit them.

The last ethical framework we'll mention here is one that emerged from our research for this book: what we call *triangulation*. It's a concept we discussed in chapter 7, in our review of crowdsourced citizen science platforms like the Global Fireball Observatory. Like the other ethical models we've discussed, it's a fairly simple idea in theory, but challenging to develop into enforceable rules or reliable practices.

The basic premise of triangulation is that AI and other algorithms are not only more just and equitable, but also more accurate and reliable, when they are built in a way that treats multiple perspectives as equally valid rather than attempting to shoehorn them into a falsely unitary vantage point (Donna Haraway's "God trick"), or privileging some contributions (and contributors) over others. In cases where two or more outlooks can't be reconciled, an algorithm doesn't necessarily have to generate an illusory consensus or render a seamless model of the world; in its design and interface, an algorithm can present versions, options, ellipses, and overlaps, giving end users the freedom and the power to navigate complexity collectively, based on their own cultural values and expertise, just as they do in the "real world."

There's no technological need to sacrifice liberty at the altar of convenience, or to present a false image of uniformity to paper over a diverse set of viewpoints, identities, and experiences. If anything, the desire for false uniformity is a cultural one, arising from historical political forces that subordinate the needs of the many to the needs of the few. And the technological sheen of algorithmic systems obscures an important fundamental truth about them: they're entirely reliant on statistical inferences, different only in scale from any other method of making informed guesses about how the world works. This is why we must dismantle the widespread myth that AI and ML are "perfect" or "know better" than humans do.[11] Instead, we need to think of algorithms as fundamentally social and communicative systems that function—and succeed—only to the extent that they mediate between people with different needs and perspectives.

While new laws and ethical frameworks are valuable ways to alter the role that tech plays in our lives, sometimes the most direct route is to change the technology itself or the way it's made and used. So for our final set of potential slow fixes, we'll focus on some promising interventions on that front. Again, this is hardly a comprehensive overview of every radical new idea in the tech arena, but rather a handful of examples that piqued our interest in light of this book's findings.

How can we make algorithms more transparent and accountable? This is one of the most widely discussed and debated questions in tech today, for good reason. As our interviewees frequently observed, the black-box nature of most AI and ML models is a problem that not only obscures dangerous biases and inaccuracies, but also provides a degree of plausible deniability for those who build and use such tech in business and society.

Some researchers and advocates, such as Mozilla fellow Deb Raji, have suggested what may be the most pragmatic solution: the development of an algorithmic auditing regime.[12] In this framework, independent third parties designated by a government body or standards organization would have regular access to algorithmic source code and would either certify that it meets certain thresholds before commercial deployment or play a role in mediating between tech developers and the individuals and communities who allege that the technology causes harm.

Other advocates have suggested a complementary approach: creating independent standards for identifying and addressing bias in AI and ML systems. In 2022, the National Institute of Standards and Technology (NIST) at the United States Department of Commerce published an extensive study laying the foundations for a system that would standardize the identification and management of AI bias. Echoing the work of advocates and scholars like Laura Forlano, NIST based its approach on what it calls a "human-centered design process for AI systems," acknowledging from the outset that algorithmic bias extends "beyond the computational level," and therefore any plan to audit and hold such systems accountable must also take into account the "socio-technical contexts" in which AI is built and used.[13]

Other approaches to algorithmic accountability address the design and implementation of the systems themselves. For instance, in 2018 a group of computer scientists at Google released a paper proposing that machine learning algorithms be released along with "model cards" that would explain in clear language and with reliable documentation precisely what their training sets comprise, how well they perform against certain benchmarks, in what social contexts the models are intended to be used, and other relevant information.[14] Google seized on the idea and published initial prototype model cards for face detection and object detection algorithms in 2020. However, a few months later, the company fired several of the paper's authors, and since then the company's model card project appears to have stalled.[15] Nonetheless, in the wake of these developments, there has been a broader push for AI and ML developers to become more open source about their training sets and social uses. One of the organizations leading this effort is Hugging Face, an AI software company that currently employs Margaret Mitchell, the former leader of Google's model cards project.

It's admirable when giant companies like Google or well-funded startups like Hugging Face set out to build more transparent and accountable algorithms. But they're still the exception rather than the rule, and most efforts to mitigate algorithmic bias and other AI-related harms come from third parties with far smaller budgets and loftier goals.

Earlier, we discussed the ethic of indigenous data sovereignty, which addresses the ways in which power structures dating back to the colonial era tend to be replicated by new data systems. One example of this ethic put into practice is the effort by indigenous coders to make AI and ML models more responsive to their own cultural systems and needs. For instance, a now defunct startup called Intelligent Voices of Wisdom (IVOW) spent several years developing what it called an Indigenous Knowledge Graph: a relational database of Native American and other folkloric information, collected and coded by members of indigenous tribes and communities, that could be integrated into larger ML and AI models to address their inherent biases and shortcomings.[16]

An analogous project is Sufi Plug Ins, a music production software suite developed by the musician Jace Clayton, who performs under the name DJ/Rupture.[17] This software intentionally challenges the Eurocentric aesthetics embedded in all of the major audio production software packages, which make it difficult to produce music that diverges from the cultural values and social expectations associated with Western stylistic conventions. Nearly everything about Sufi Plug Ins calls European musical norms into question. In addition to the twelve notes of the piano keyboard, the software includes synthesizers tuned to North African maqam scales; controls are labeled by default in Berber script instead of English; and an optional component of the software lowers the users' computer volume five times a day to accommodate the Muslim call to prayer.

In addition to empowering users to create music that sounds different from European classical or Western pop music, Clayton's software empowers musicians to critique the industry-standard software they're accustomed to using, by highlighting the ways in which its default settings and functions are specific to Western cultural values and expectations. In this project, data sovereignty isn't merely conceptualized in terms of freedom from surveillance and data extraction; it's also framed in terms of freedom from constraint in media production. Or, to put it a bit differently: sovereignty and justice in a networked society aren't just about being free *from* data, but also being free *with* data.

SLOWER CHALLENGES

Despite all the legal, ethical, and technological interventions we've discussed, the secret life of data still poses some very thorny problems for culture and society, some of which may be so resistant to change that they outlast many of the proposed solutions. For one thing, even though there's a long list of well-documented challenges and innumerable good ideas about how to address them, implementing these changes would still require a level of dedicated funding, political will, and institutional coordination that doesn't currently exist. As we've mentioned, there has been no binding law, regulation, or treaty anywhere asserting broad affirmative human rights with respect to algorithms. Before any such proposal is even introduced, it inevitably contains numerous exemptions and concessions calculated to mollify data industries, law enforcement, and other organizations that benefit from the unchecked ability to surveil, predict, and influence the behavior of citizens and consumers through data collection and analysis. And even with these concessions, enacting such a proposal is still, at best, a remote possibility that would require years of coordinated political maneuvering.

The lack of adequate funding is also a major challenge. Google's model cards, the Indigenous Knowledge Graph, and Sufi Plug Ins are now defunct projects (though Mitchell has continued the model cards project at Hugging Face). As IVOW founder Davar Ardalan wrote when she announced the shuttering of her entire organization in 2022, "We wish we had funding to help create these knowledge engines responsibly but alas we don't."[18] She laid the blame for this lack of support directly at the feet of those who finance big tech, especially venture capitalists (VCs), and predicted that their narrow worldview and agenda would have severe consequences far beyond the fate of her own project. "There is no willingness to pause and think of beneficial AI," she wrote. "The lack of vision on the part of VCs and investors will cripple the development of AI."[19]

In the meantime, most of the supposedly privacy-preserving innovations being funded and adopted by big investors and major software developers fall short when it comes to combating the social consequences of surveillance. For instance, a flight information display screen unveiled in 2022 at the Detroit Metropolitan Airport uses a new technology called Parallel Reality to identify individual travelers and show them personalized flight information, even when hundreds are looking at the same screen at the same time. Delta Air Lines, which installed the screen, claims the "private viewing zone" constructed for each user relies on "anonymous," "non-biometric" technology.[20] Even taking into account the potential for deanonymization we discussed in chapter 3, Delta's narrow focus on the *nature* of the data ignores the sociotechnical *context* in which it is collected and used (to echo NIST's concerns). Technology like this risks not only normalizing and institutionalizing individual location tracking to a greater degree, but also alters our experience of common physical space, threatening to place us in the same kind of information silos that have damaged our civic institutions in the online context.

There's nothing particularly sinister about Delta's efforts or Parallel Reality technology. If anything, it's laudable that the company attempted to preserve privacy by providing a customized interface that doesn't rely on facial recognition technology. The problem is that this is a circle that can't be squared. As we discussed in our conversation with cybersecurity researcher Vasilios Mavroudis in chapter 7, usually, the more user friendly a system is, the worse it is at protecting privacy—and vice versa.

We also interviewed Yves-Alexandre de Montjoye, a professor who focuses on computational privacy at Imperial College London. As he explained, this tension is an inevitable feature of the architecture of large-scale networks. His research has shown that if you can get some basic information about the connections between a tiny fraction of individuals in a network, you can use it to build a surprisingly reliable map of the relationships throughout the entire web of users—like looking into a shard of a hologram and seeing the complete image reproduced in it. Montjoye's research has proven mathematically that this holographic effect is intrinsic to networks themselves and therefore can't be eradicated, only mitigated through clever design or robust regulation.[21] Therefore, he told us, "What we need as a society is to find a reasonable band" of privacy—a Goldilocks zone between too cold (in which we lose all the benefits of computational technology) and too hot (in which we lose all the benefits of privacy and data security).

The problem with that approach, of course, is that reasonable minds may differ. As anyone who's ever battled over the AC controls in a car knows, one person's too hot can be another person's too cold. This is part of the reason why the concept of consent is so fraught when it comes to data privacy policy.

If you've spent any time on the web in the 2020s, you've probably encountered a *cookie wall*—one of those pop-up alerts that asks you to opt in or out of various tracking technologies before you can get to the page you want to visit. Cookie walls sprang up all over the web as a result of the GDPR, which requires that advertisers get users' consent before collecting certain kinds of data. But is it really consent if you're not allowed to access the website for your bank, school, employer, or doctor until you've agreed to be tracked? For most internet users, it's just an annoyance, and they'll agree to virtually anything (without reading the fine print) in order to stay solvent, employed, or healthy. But for others, it can literally be a question of life and death. A nonprofit suicide hotline called Crisis Text Line drew widespread criticism in 2022 when Politico revealed that it was selling "anonymized" data from its communications with emotionally vulnerable minors and others to a for-profit customer support platform called Loris.[22] Though both companies emphasized that such uses were included under the fifty-paragraph terms of service (TOS) presented to hotline callers, few reasonable people would argue that such callers were psychologically equipped to absorb and consider the full implications of this TOS before agreeing.

Cookie walls and unreadable TOS make consent a questionable premise, but at least they place the decision nominally in the hands of those being surveilled. Yet structural privacy challenges like those described by Mavroudis and Montjoye make consent even more difficult to conceptualize and ensure by removing the targets of data collection and analysis from the decision-making process entirely. As we discussed in chapter 8, Cambridge Analytica was able to retrieve sensitive data about nearly a hundred million Facebook users based on the ostensibly voluntary participation of a few hundred thousand quiz-takers. Should those quiz-takers have the power to consent to surveillance (let alone the subsequent analysis and potential abuse of the resulting data) on behalf of all their other Facebook "friends"?

Similarly, as we discussed in chapter 5, DNA-based ancestry websites have unearthed countless family secrets that would otherwise have remained hidden. Should an internet user researching their own genealogy be granted the sole power, discretion, and responsibility to reveal, even unintentionally, that their father or grandfather had sired unacknowledged children? Until we begin to grapple with these kinds of questions, we can never arrive at a social consensus about the meaning of consent in a data-rich society. And, therefore, we can't develop laws, regulations, and technologies that live up to our expectations about justice, sovereignty, and ethical business practices.

These are just a handful of the "slower" challenges we face in trying to shape and contain the transformative role that data collection and processing play in our society. There are countless more we could add to the list, but ultimately, they all converge on the same point: we are woefully underprepared for a future—or even a present—in which global-scale sensor nets, biometric implants, IoT-connected smart devices, AI-derived predictive models, ML-based content generators, deepfakes, chatbots, GANs, Easter eggs, and malware permeate our bodies, intimate relationships, public spheres, places of business, and political institutions.

Despite the common clichés, data isn't a resource like oil, or water, or even air. It can't be contained, channeled, or monopolized. It's not subject to scarcity, and it has no intrinsic economic value. Data isn't naturally occurring; it's a byproduct. It's something we create when we try to make sense of the worlds around and within us, and its contours and flows are a map of the technological toolsets, social relationships, cultural symbolic systems, and political power dynamics that already exist in those worlds. But, like other byproducts, such as greenhouse gasses and nuclear waste, data may also warp the environment that shaped it, creating feedback loops between input and output that can accelerate rapidly toward unpredictable and extreme outcomes.

We can rarely guess, let alone be certain, what those outcomes might be, because the only thing we know for sure about data is that it is bound to have a secret life. Whatever its purposes, functions, or meanings may have been when it was first collected, analyzed, or sold, it will inevitably be used for radically different purposes in radically different contexts. Water may always flow toward the sea, seeking the earth's center of gravity, but (if you'll permit us the metaphorical flight of fancy) data does the opposite: it spreads out, away from its source, escaping the gravity well of its origins and cascading through the boundless space beyond, where it may be collected and used by unknown recipients on innumerable strange and distant worlds.

We can't control data, but we can understand it better. We *must* understand it better if we hope to retain the power to shape our identities, our cultures, and our democracies as information technology grows stronger and more ubiquitous with every passing year. That means we must stop focusing reactively on its first-order consequences and instead develop an affirmative conceptual language to think through its second-, third-, and nth-order implications. We must stop treating AI and ML algorithms as autonomous, impenetrable black boxes, and we must instead hold the human beings who build, finance, and use such algorithms responsible for their inner workings and their outcomes. We must stop treating science fiction as a blueprint for the future and instead rely on it as a guide to potential pitfalls. We must center justice in our designs, our regulations, and our markets. Most importantly, we must remember that technology—no matter how "smart"—is built by and for people, and we must wield it together for the creation of collective knowledge, not the abuse of concentrated power.

POSTSCRIPT: THE SHAKE OF THINGS TO COME

Concluding a book on technology and society is kind of like trying to land a 747 on a moving treadmill in an earthquake. Even if we pull it off perfectly (we haven't), we're still hitting the ground faster than we can handle, in the midst of major upheaval. The technosocial landscape has shifted considerably during the thirty months we spent writing and editing this book, and by the time you read these words, it's sure to have changed even more.

But even though we've just been talking about how uncertain the future is, we'd be remiss in our duties if we didn't point out some big, blurry shapes on the horizon and call attention to the ways in which they might betoken even bigger changes with profound implications for the secret life of data. We'll keep it short and sweet (and overly simple) for now, but we have a feeling we'll have to spend a lot more time researching and writing about some of these issues in the years to come.

First of all, networked society may be headed for a disaster that some cryptology researchers refer to as the *quantum cliff*. Computer encryption (which, as we've discussed, is vital to password-protected apps and services, virtual private networks, secure chat protocols, safe online banking, private medical records, and protecting state secrets) depends on the fact that it's mathematically easier to scramble information than to unscramble it. Your laptop *could* eventually unscramble your neighbor's financial records or secret nuclear weapon plans on government servers, but it would take gazillions of years to calculate, so it doesn't really matter. However, quantum computing, which relies on the weirdness of subatomic physics to accelerate information processing speed drastically, is developing at a rapid pace. And, while quantum computing is currently too slow and too costly to be available widely to consumers, researchers estimate that by 2030, quantum computers will be fast enough to render most encryption used today as useless as a cardboard padlock.[23]

That's why, since the mid-2010s, information security researchers have focused on developing something called *quantum encryption*, which should render information just as secure in a quantum computing environment as traditionally encrypted information has been in our current digital computing environment. Unfortunately, it's a race against time; in 2022, NIST announced plans to develop standards for quantum encryption by 2024, with the goal of "mitigating as much of the quantum risk as is feasible by 2035."[24] Not exactly reassuring words, especially for those whose lives and livelihoods depend on keeping information secure.

Even more troubling is the fact that none of the traditionally encrypted information out there can be secured retroactively by quantum encryption. At this moment, countless state, commercial, and criminal organizations are intercepting as much encrypted internet traffic as they can and storing it on massive servers, waiting for the day, not long from now, when its secrets can be unlocked cheaply and easily via quantum decryption.

What this means from a practical standpoint is that, at some point in the next decade or two, these actors may be able to access, analyze, and search through archives of anything you've ever said, done, transacted, or recorded using a networked computing device (including your smartphone, laptop, smart appliance, and biometric tracker), as well as many of the database entries analyzing your behavior and identity compiled by data brokers, government organizations, insurance companies, credit bureaus, and others. And, if history is any guide, it's entirely possible that large swaths of this data and analysis will eventually make their way into the public eye, where they will be more or less available for any interested parties to review in perpetuity. Yes, you read that correctly.

The quantum cliff is a somewhat likely near-term scenario—likely enough that the White House issued an extensive National Security Memorandum in 2022 outlining quantum computing leadership as a national priority and security vulnerability.[25] But further developments in both quantum and traditional computing may also contribute to less predictable and even more potentially transformative scenarios in the not-so-distant future.

One of the biggest unknowns is whether and how AI will develop beyond its current, highly focused role as a computational tool and into something more autonomous or even self-aware. Computer scientists are divided on this question, with many claiming that "general AI" (human-level intelligence or greater) is an impossibility, others claiming that it's a far-off possibility, and a handful claiming it's already upon us. For instance, in 2022, a software engineer at Google named Blake Lemoine briefly made headlines when he published logs of a conversation he and another engineer had with a proprietary Google chatbot called LaMDA, which he claimed demonstrated that the software program was sentient.[26]

Though the logs make fascinating reading, and could understandably lead a reader to believe that LaMDA was attempting to communicate self-awareness, Google leadership and many unaffiliated computer scientists quickly stepped up to dispute Lemoine's claims, arguing that, like the legendary Pygmalion, he'd fallen so deeply in love with his own creation that he could no longer see the artifice behind it. Google promptly fired Lemoine, but not before his leak had reignited longstanding debates over whether AI sentience is possible, and what an appropriate human response should be if it is.[27]

In the meantime, Lemoine is far from the only one arguing that AI is on track to radically alter the human condition in the foreseeable future. Whether or not it achieves something people would recognize as sentience, many experts argue, AI is likely to gain a degree of power and complexity that will challenge human autonomy and cause major disruptions to our way of life, or even end it. For example, historian and philosopher Émile P. Torres penned an op-ed in the *Washington Post* in 2022 arguing that "artificial superintelligence" poses an existential threat to life on earth—not out of malice, but "on accident," because it would prioritize its own directives over human survival.[28] Along very similar lines and in the same year, computer scientists from Oxford and Australian National University published a paper in the academic journal *AI Magazine* arguing that a "sufficiently advanced artificial agent" could essentially hack human behavior, leading to "catastrophic consequences."[29] And engineer Max Hodak, cofounder of brain-computer interface company Neuralink, made an analogous argument on Twitter in 2021, bluntly concluding that "we are going to get so wrecked" by AI, because "machines might end up having a lot more flexibility on how they organize themselves than we do."[30]

We are certainly not claiming that the AI apocalypse is nigh. As we've taken pains to point out throughout this book, many of the grandest claims made for computer intelligence are unsupported by concrete evidence, and there's a lot of smoke and mirrors behind even the most successful and functional programs. For instance, Pomona College economist Gary Smith, whose work focuses on debunking overblown AI rhetoric, has criticized machine learning language model GPT-3, arguing that its developer "OpenAI evidently employs 40 humans to clean up GPT-3's answers manually because GPT-3 does not know anything about the real world."[31]

Ultimately, it may not really matter whether LaMDA is sentient or not, or whether GPT-3 pulls off its feats of lexical dexterity with an assist from human contributors. Either way, the geometrically accelerating speed of information processors (especially quantum computers), combined with the rapid integration of algorithms into every aspect of our lives and communities, means that the secret life of data will continue to expand in ways we can't predict or control.

We may be obliterated by superintelligent AIs that treat us with the same wanton disregard we show to the other species on our planet. We may engineer ourselves out of jobs, homes, and health in the name of progress and profit. We may successfully fuse human and machine intelligence, ushering in a new posthuman era. We may develop new political philosophies and collaborative systems that exploit the power of algorithms to engineer a more just, equitable, and stable society. We may, as futurist Ray Kurzweil has long argued, be approaching a singularity in which we transfer our consciousnesses into a giant computer server, living freely and happily in the digital ether until the heat death of the universe billions of years from now.[32] Who knows? We certainly don't, and neither does anyone else. But the least likely scenario of all is stasis—the preservation of a human society unaltered by rampant data collection and untouched by the power of computation.

We're not trying to be alarmist. If we've left you paralyzed with anxiety, depression, or anger at the fate of the world, then we've failed in our basic duty. What we hope you take from this book is the same thing we've gotten from writing it: a sense of awe, humility, and wonder at the power of information, a conceptual scaffolding to help make sense of the disjointed narratives about technology and society swirling through the media, and most importantly, a renewed sense of empowerment and purpose. Your future is at stake, and therefore your voice is essential to this conversation.

It's important to remember that the future isn't a fixed point, and we're not captives on some carnival ride, hurtling toward our destination on a preordained track. The past is likewise still a mystery; data has always had a secret life, and the more we discover about the universe we live in, the more likely we are to find its echoes imprinted on the fabric of the cosmos, from the quantum realm to the building blocks of life to the filaments of dark matter stringing galaxies like tinsel across the span of light-years.

The sheer scale of it all can be overwhelming—too much to fit into one brain, or one lifetime. But you're not alone. We're all part of a global network spanning centuries and evolving by the millisecond. And right here, right now, we can all help shape what comes next. We can continue down the path of least resistance and allow our lives to be dictated by a handful of powerful interests designing secretive systems that lack accountability, let alone consciences. Or, we can use our powerful new information processing tools to learn more about ourselves, one another, and the world around us, gaining the wisdom to draw conclusions from our past so that we might create a better future together.

Acknowledgments

This book is the result of two and a half years of sustained research, collaboration, writing, and editing, conducted collaboratively by the coauthors via Zoom, Slack, Zotero, Google Drive, FaceTime, email, and SMS at a distance of roughly 2,600 miles. Its genesis was a collaboratively written article that required a year of labor, which was followed by a year of preparing and pitching a book proposal. All in all, by the time these words appear in print, we will have spent roughly six years total on them from inception to completion. We could not have made it a month, let alone the better part of a decade (which also contained a global pandemic), without the significant intellectual, emotional, and material support offered by our loved ones, our colleagues, our expert sources, our editors, and our home institutions.

First and foremost, we'd like to thank the people who kept us relatively sane and healthy throughout these years. For Jesse, they are my fiancée, Mona Tian; my parents, Lucy Gilbert, Cornelia Brunner, Roberta Sklar, and Sondra Segal; my sister, Rachel Segal-Sklar; extended family members Lela Zaphiropoulos and Jan van Assen; and my dear friends Chauncy Godwin, Miguel Atwood-Ferguson, Ahmed Best, Aya Peard, Simon Overbey, Scott Rosenberg, and Pablo Molina. For Aram, they are my wife, Dunia Best Sinnreich; our children, Simon and Asha; my parents, Masha Zager, Jonathan Sinnreich, and Emily Pines; my in-laws, Bahati and Ahmondylla Best; my siblings, Rachel Cleves and Dan Sinnreich; and dear friends Henry Myers, Patricia Aufderheide, Vivien Goldman, Nicholas John, Joel Gershon, and Noah Shachtman. Many of these people also served as sounding boards for our ideas, early editors of our draft chapters, and de facto therapists whenever the gravity of our subject matter weighed too heavily on our hearts. Our debt to you is incalculable, and we hope that this book is worthy of all you've given us.

Other friends and colleagues offered additional support and feedback. Jonathan Sterne deanonymized himself as a peer reviewer when we first submitted "The Carrier Wave Principle" to the *International Journal of Communication*. His brilliant and constructive comments pushed us to approach the subject with rigor, precision, and clarity. Laura DeNardis was an early and enthusiastic supporter of this project, and her triple role as cheerleader, book series coeditor, and expert source has been invaluable. Special thanks to Neil W. Perry from American University, who provided key support to us in preparing the manuscript for publication, meeting our hectic deadlines and strenuous demands with good humor and professionalism. Jesse would like to thank Cornelia Brunner for her insights and encouragement, and for her lifelong mentorship. I owe much of who I am and how I think about the world to you. Aram would also like to thank his scholarly colleagues at the American University School of Communication and at the Association of Internet Researchers. It's an honor to work among such smart, ethical, forward-thinking people, and this book reflects much of what I've learned from you.

We acknowledge and thank the dozens of experts we spoke to throughout this project, who provided a window into their practices and challenged our assumptions about technology and culture in ways we could never have anticipated. Without your brilliant contributions, this book would not exist.

Finally, we would like to thank the staff at the MIT Press, especially our editor, Justin Kehoe, who has been supportive, critical, hands-on, and hands-off in exactly the right proportions throughout the process from book pitch to first draft to peer review to final revision and beyond. We couldn't ask for a better home for this project, and we hope to work with you again the next time around.

Notes

INTRODUCTION

1. Justin Jouvenal, Mark Berman, Drew Harwell, and Tom Jackman, "Data on a Genealogy Site Led Police to the 'Golden State Killer' Suspect. Now Others Worry about a 'Treasure Trove of Data,'" *Washington Post*, April 27, 2018, https://www.washingtonpost.com/news/post-nation/wp/2018/04/27/data-on-a-genealogy-site-led-police-to-the-golden-state-killer-suspect-now-others-worry-about-a-treasure-trove-of-data.

2. Ashley May, "Took an Ancestry DNA Test? You Might Be a 'Genetic Informant' Unleashing Secrets about Your Relatives," *USA Today*, April 27, 2018, https://www.usatoday.com/story/tech/nation-now/2018/04/27/ancestry-genealogy-dna-test-privacy-golden-state-killer/557263002.

3. Paige St. John, "The Untold Story of How the Golden State Killer Was Found," *Los Angeles Times*, December 8, 2020, https://www.latimes.com/california/story/2020-12-08/man-in-the-window.

4. St. John, "Untold Story."

5. Technically, Dobbs v. Jackson Women's Health Organization, 142 S. Ct. 2228 (2022).

6. According to Pew Research surveys, in 2022, 61 percent of US adults believed abortion should be legal in all or most cases. This was the highest percentage reported in twenty-seven years of annual polling on the issue.

7. We first explored this concept in a 2019 academic journal article, where we referred to it as the "carrier wave principle." If you're interested in a deeper dive into the scholarly terrain surrounding and informing our analysis in this book, we encourage you to read it as a companion piece. See Aram Sinnreich and Jesse Gilbert, "The Carrier Wave Principle," *International Journal of Communication* 13 (2019): 5816–5840.

8. *Data* is technically a plural, and we treat it as such by default, but there are contexts (including this book's title) in which we choose to treat it as a singular noun instead.

9. Aram Sinnreich, Patricia Aufderheide, Maggie Clifford, and Saif Shahin, "Access Shrugged: The Decline of the Copyleft and the Rise of Utilitarian Openness," *New Media and Society* 23, no. 12 (2020): 3466–3490; Fernando R. Rosa, Maggie Clifford, and Aram Sinnreich, "The More Things Change: Who Gets Left Behind as Remix Goes Mainstream?" in *The Routledge Handbook of Remix Studies and Digital Humanities*, eds. E. Navas, O. Gallagher, and X. Burrough (New York: Routledge, 2021), 36–52.

CHAPTER 1

1. Dani Deahl, "Metadata Is the Biggest Little Problem Plaguing the Music Industry," *The Verge*, May 25, 2019, https://www.theverge.com/2019/5/29/18531476/music-industry -song-royalties-metadata-credit-problems.

2. Jorge Luis Borges, *The Library of Babel* (Boston: David R. Godine; Enfield, UK: Airlift, 2000).

3. Barton Gellman, "Inside the NSA's Secret Tool for Mapping Your Social Network," *Wired*, February 26, 2021, https://www.wired.com/story/inside-the-nsas-secret-tool-for-mapping -your-social-network.

4. Pew Research, "Obama's NSA Speech Has Little Impact on Skeptical Public," *Pew Research Center*, January 2014, https://www.pewresearch.org/wp-content/uploads/sites/4/legacy -pdf/1-20-14-NSA-Release.pdf.

5. Jonathan Mayer, Patrick Mutchler, and John C. Mitchell, "Evaluating the Privacy Properties of Telephone Metadata," *Proceedings of the National Academy of Sciences* 113, no. 20 (May 2016), https://www.pnas.org/content/113/20/5536.

6. Gellman, "Inside the NSA."

7. Xeni Jardin, "P2P in the Legal Crosshairs," *Wired*, March 15, 2004, https://www.wired .com/2004/03/p2p-in-the-legal-crosshairs.

8. Erika Eichelberger, "Lobbyist Secretly Wrote House Dems' Letter Urging Weaker Investor Protections," *Mother Jones*, August 30, 2013, https://www.motherjones.com/politics/2013 /08/congressional-black-caucus-fiduciary-duty-rule-financial-services-institute.

9. Alex Rogers, "House Introduces Online Gambling Bill Backed by Sheldon Adelson," *Time*, February 4, 2015, https://time.com/3695948/sheldon-adelson-online-gambling.

10. Dustin Weaver, "Adelson Finds Allies in Gambling Crusade," *The Hill*, March 20, 2014, https://thehill.com/business-a-lobbying/business-a-lobbying/201289-adelson-finds-allies -on-gambling-ban.

11. Graham Cluley, "Fugitive John McAfee's Location Revealed by Photo Meta-Data Screw-Up," *Naked Security* (blog), December 3, 2012, https://nakedsecurity.sophos.com/2012/12/03 /john-mcafee-location-exif.

12. Cluley, "Fugitive John McAfee's Location Revealed."

13. Cluley, "Fugitive John McAfee's Location Revealed."

14. McAfee died in a prison cell in Barcelona in 2021, shortly before he was due for extradition back to the United States on tax evasion charges unrelated to the Faull case.

15. Winner was released early from prison to a transitional facility by President Biden in 2021, but her conviction stands at the time of writing.

16. Matthew Cole, Richard Esposito, Sam Biddle, and Ryan Grim, "Top-Secret NSA Report Details Russian Hacking Effort Days before 2016 Election," *The Intercept*, June 5, 2017, https://theintercept.com/2017/06/05/top-secret-nsa-report-details-russian-hacking-effort -days-before-2016-election.

17. Electronic Frontier Foundation, "List of Printers Which Do or Do Not Display Tracking Dots," n.d., https://www.eff.org/pages/list-printers-which-do-or-do-not-display-tracking -dots. Incidentally, your laser printer probably uses tracking dots, as well. We highly recommend you try reproducing our exercise to see what kind of invisible metadata you're sharing with each page you print.

18. Reality Winner was still imprisoned when we conducted this interview.

19. Technically, it's the company's predictive algorithm, not the data itself, that becomes more accurate.

20. Todd Spangler, "Netflix Execs Defend Cancellations, Saying 93% of Series Have Been Renewed," *Variety*, July 17, 2017, https://variety.com/2017/digital/news/netflix -cancellations-original-series-renewals-1202497938.

21. Bruce Schneier, "Metadata = Surveillance," *Schneier on Security* (blog), March 13, 2014, https://www.schneier.com/blog/archives/2014/03/metadata_survei.html.

22. Laura DeNardis, *The Internet in Everything: Freedom and Security in a World with No Off Switch* (New Haven: Yale University Press, 2020).

23. Aaron Sankin, "Want to Find a Misinformed Public? Facebook's Already Done It," *The Markup*, April 23, 2020, https://themarkup.org/coronavirus/2020/04/23/want-to-find -a-misinformed-public-facebooks-already-done-it.

CHAPTER 2

1. Dominique Aubert-Marson, "Sir Francis Galton: The Father of Eugenics," *Medicine Sciences* 25, nos. 6–7 (2009): 641–645.

2. Francis Galton, *Memories of My Life* (London: Methuen, 1908).

3. R. Yip, "Altitude and Birth Weight," *The Journal of Pediatrics* 111, no. 6 (1987): 869–876; NASA's Scientific Visualization Studio, "Shifting Distribution of Land Temperature Anomalies, 1951–2020," April 23, 2021, https://svs.gsfc.nasa.gov/4891; Gerardo Segura, Patricia Balvanera, Elvira Durán, and Alfredo Pérez, "Tree Community Structure and Stem Mortality along a Water Availability Gradient in a Mexican Tropical Dry Forest," *Plant Ecology* 169 (2002): 259–271; L. O. Tedeschi, W. Chalupa, E. Janczewski, D. G. Fox, C. Sniffen, R. Munson, P. J. Kononoff, et al., "Evaluation and Application of the CPM Dairy Nutrition

Model," *Journal of Agricultural Science* 146, no. 2 (2008): 171–182; K. L. S. Gunn and J. S. Marshall, "The Distribution with Size of Aggregate Snowflakes," *Journal of Atmospheric Sciences* 15, no. 5 (1958): 452–461.

4.   Harry Bruinius, *Better for All the World: The Secret History of Forced Sterilization and America's Quest for Racial Purity* (New York: Vintage, 2007).

5.   Nora McGreevey, "College Sophomores Discover Hidden Text in Medieval Manuscript," *Smithsonian Magazine*, November 24, 2020, https://www.smithsonianmag.com/smart-news /students-discover-hidden-text-medieval-manuscript-180976385.

6.   Harry Seymour, "'The Most Beautiful Female Bust' Brought Back to Life with Vivid Colour," *Art Newspaper*, September 23, 2019, http://www.theartnewspaper.com/news /beauty-brought-back-to-life-in-vivid-colour.

7.   Then again, *they might not*. These models are probabilistic in nature, which means there's always a chance that they get the details wrong, and they reflect the assumptions and biases of their creators rather than the historical truth of the artifacts being augmented. You can view Brøns's digital reconstruction here: https://www.theartnewspaper.com/2019/09/23 /the-most-beautiful-female-bust-brought-back-to-life-with-vivid-colour.

8.   "Audio Preservation with IRENE," Northeast Document Conservation Center, n.d., https:// www.nedcc.org/audio-preservation/irene.

9.   Karen Trentelman, Koen Janssens, Geert van der Snickt, Yvonne Szafran, Atte T. Wool-lett, and Joris Dik, "Rembrandt's *An Old Man in Military Costume*: The Underlying Image Re-examined," *Applied Physics A* 121 (2015): 801–811, https://doi.org/10.1007 /s00339-015-9426-3.

10. Meilan Solly, "Researchers Uncover Hidden Details Beneath Picasso Painting," *Smithsonian Magazine*, February 19, 2018, https://www.smithsonianmag.com/smart-news /researchers-uncover-hidden-details-beneath-picasso-painting-180968204.

11. Marisa Iati, "A Message in 'The Scream' Stumped Historians for Years. Now They Think They Know the 'Madman,'" *Washington Post*, February 23, 2021, https://www.washingtonpost .com/history/2021/02/23/scream-painting-inscription.

12. Mark Glickman, Jason Brown, and Ryan Song. "(A) Data in the Life: Authorship Attribution in Lennon-McCartney Songs," *Harvard Data Science Review* 1, no. 1 (July 2019), https:// doi.org/10.1162/99608f92.130f856e.

13. William J. Broad, "New Technique Reveals Centuries of Secrets in Locked Letters," *New York Times*, March 2, 2021, https://www.nytimes.com/2021/03/02/science/locked-letters -unfolding.html.

14. Shannon Sims, "In Brazil, a New Rendering of a Literary Giant Makes Waves," *New York Times*, June 14, 2019, https://www.nytimes.com/2019/06/14/books/brazil-machado-de -assis.html.

15. Aram Sinnreich, *Mashed Up: Music, Technology, and the Rise of Configurable Culture* (Amherst: University of Massachusetts Press, 2010).

16. Kyle McDonald, *Light Leaks*, filmed November 9–10, 2019, by Fulcrum Arts in Pasadena, CA, video, 11:14, https://www.youtube.com/watch?v=mONzmKPlDFo.

17. Jeremy Hsu, "Smartphone Cameras Peek around Corners by Analyzing Patterns of Light," *IEEE Spectrum*, October 9, 2017, https://spectrum.ieee.org/mit-shows-how-smartphones -could-peek-around-corners; Terry Devitt, "Lessons of Conventional Imaging Let Scientists See around Corners," *University of Wisconsin News*, August 5, 2019, https://news.wisc.edu /lessons-of-conventional-imaging-let-scientists-see-around-corners.

18. H. Y. Wu, M. Rubinstein, E. Shih, J. Guttag, F. Durand, and W. Freeman, "Eulerian Video Magnification for Revealing Subtle Changes in the World," *ACM Transactions on Graphics* 31, no. 4: 121–130, 2012.

19. The GDELT Project, https://blog.gdeltproject.org/gdelt-translingual-translating-the-planet.

20. Masahiro Mori, "The Uncanny Valley," *Energy* 7, no. 4 (2012): 33–35. [Originally published in 1970, in Japanese.]

21. As you may have noticed, fooling humans is an important concept in AI. We will discuss the implications of this further in the pages to come.

22. Arthur C. Clarke, *Profiles of the Future: An Enquiry into the Limits of the Possible* (London: Victor Gollancz, 1962).

23. Keith Seward, "Through the Looking Glass," *Artforum* 31 (October 1992): 107. https://www .artforum.com/print/reviews/199208/through-the-looking-glass-57282.

24. Lumière Films, *Arrival of a Train at La Ciotat* (1896), video, at 00:52, https://archive.org /details/11-arrival-of-a-train-at-la-ciotat.

25. Marshall McLuhan, *Understanding Media: The Extensions of Man* (Canada: McGraw-Hill, 1962).

26. Scott McCloud, *Understanding Comics: The Invisible Art* (New York: William Morrow, 1994).

27. Tom Brown, Benjamin Mann, Nick Ryder, Melanie Subbiah, Jared Kaplan, Prafulla Dhari-wal, Arvind Neelakantan, et al, "Language Models are Few-Shot Learners," *Advances in Neural Information Processing Systems* 33 (2020): 1877–1901, https://proceedings.neurips .cc/paper/2020/file/1457c0d6bfcb4967418bfb8ac142f64a-Paper.pdf.

28. "Write with Transformer," Hugging Face, https://transformer.huggingface.co/doc/gpt2-large.

29. E. M. Bender, T. Gebru, A. McMillan-Major, M. Mitchell, V. Prabhakaran, M. Díaz, and B. Hutchinson, "On the Dangers of Stochastic Parrots: Can Language Models Be Too Big?," in *Proceedings of the 2021 ACM Conference on Fairness, Accountability, and Transparency* (New York: ACM, 2021), 610–623.

30. Jill Abramson, *Merchants of Truth: The Business of News and the Fight for Facts* (New York: Simon and Schuster, 2019).

31. Michael C. Moynihan (@mcmoynihan), Twitter, February 6, 2019, https://twitter.com/mcmoynihan/status/1093290016632115202?lang=en.

32. *All Things Considered*, "Former 'NYT' Executive Editor Jill Abramson Responds to Plagiarism Allegations," NPR, February 7, 2019, https://www.npr.org/2019/02/07/692466442/former-nyt-executive-editor-jill-abramson-responds-to-plagiarism-allegations.

33. At the time of writing, a Google search for this full sentence yielded zero results.

34. Michael R. Sisak, "Modern Policing: Algorithm Helps NYPD Spot Crime Patterns," Associated Press, April 20, 2021, https://apnews.com/article/police-us-news-ap-top-news-artificial-intelligence-crime-84fb03384368458db3d85763b5bf5b94.

35. Andrea, Peterson, "LOVEINT: When NSA Officers Use Their Spying Power on Love Interests," *Washington Post*, August 24, 2013, https://www.washingtonpost.com/news/the-switch/wp/2013/08/24/loveint-when-nsa-officers-use-their-spying-power-on-love-interests; Ed Pilkington, "NYPD Settles Lawsuit after Illegally Spying on Muslims," *The Guardian*, April 5, 2018, https://www.theguardian.com/world/2018/apr/05/nypd-muslim-surveillance-settlement.

36. Drew Harwell and Eva Dou, "Huawei Tested AI Software That Could Recognize Uighur Minorities and Alert Police, Report Says," *Washington Post*, December 8, 2020, https://www.washingtonpost.com/technology/2020/12/08/huawei-tested-ai-software-that-could-recognize-uighur-minorities-alert-police-report-says.

37. Nick Waters, "The Killing of Muhammad Gulzar," Bellingcat, May 8, 2020, https://www.bellingcat.com/news/uk-and-europe/2020/05/08/the-killing-of-muhammad-gulzar; Bushra Shakhshir, and Bulent Usta, "Turkey Says Greek Forces Kill Migrant at Border, Athens Denies Claim," *Reuters*, March 4, 2020, https://www.reuters.com/article/uk-syria-security-turkey-greece-migrants/turkey-says-greek-forces-kill-migrant-at-border-athens-denies-claim-idUKKBN20R23V.

38. Waters, "Killing of Muhammad Gulzar."

39. Matt Korda, "Widespread Blurring of Satellite Images Reveals Secret Facilities," *Federation of American Scientists* (blog), https://fas.org/blogs/security/2018/12/widespread-blurring-of-satellite-images-reveals-secret-facilities.

40. danah boyd and Michael Golebiewski, "Data Voids: Where Missing Data Can Easily Be Exploited," *Data and Society*, November 2019, https://datasociety.net/wp-content/uploads/2019/11/Data-Voids-2.0-Final.pdf.

41. In fact, things have changed even less than we might think; Facebook founder Mark Zuckerberg notoriously started his career by coding a website called Facemash, which allowed users to rank the attractiveness of female students at Harvard, where he was an undergraduate.

CHAPTER 3

1. Aram Sinnreich, *The Essential Guide to Intellectual Property* (New Haven: Yale University Press, 2019).

2. As Nakamura has argued, the presumption of online anonymity was always a tacit assertion of white privilege and was therefore not experienced by internet users of color in the same way. See Lisa Nakamura, *Cybertypes: Race, Ethnicity, and Identity on the Internet* (New York: Routledge, 2002).

3. Samarati Pierangela and Latanya Sweeney, "Protecting Privacy When Disclosing Information: k-anonymity and Its Enforcement through Generalization and Suppression," Technical Report SRI-CSL-98-04 (Computer Science Laboratory, SRI International, 1998), https:// epic.org/wp-content/uploads/privacy/reidentification/Samarati_Sweeney_paper.pdf.

4. US Department of Health and Human Services, "Standards for Privacy of Individually Identifiable Health Information," 65 Fed. Reg. 82462 (December 28, 2000).

5. The Pillar, "Pillar Investigates: USCCB Gen Sec Burrill Resigns after Sexual Misconduct Allegations," *The Pillar*, July 20, 2021, https://www.pillarcatholic.com/pillar-investigates -usccb-gen-sec.

6. Dan Goodin, "Dear Ashley Madison User, I Know Everything about You. Pay up or Else," *Ars Technica*, February 1, 2020, https://arstechnica.com/information-technology/2020/02 /four-plus-years-later-ashley-madison-hack-is-used-in-new-extortion-scam.

7. Hayley Tsukayama, "Anonymous Posts File Claiming to Have Information from 4,000 Bank Execs," *Washington Post*, February 5, 2013, https://www.washingtonpost.com/business /technology/anonymous-posts-file-claiming-to-have-information-from-4000-bank-execs /2013/02/05/8f8b0488-6f9b-11e2-ac36-3d8d9dcaa2e2_story.html.

8. Roland Barthes, "The Death of the Author," in *Image, Music, Text*, trans. Stephen Heath (New York: Hill and Wang, 1977), 142–148.

9. Aram Sinnreich, *Mashed Up: Music, Technology, and the Rise of Configurable Culture* (Amherst: University of Massachusetts Press, 2010), http://www.jstor.org/stable/j.ctt5vk8c2.

10. Michel Foucault, *The History of Sexuality: The Will to Knowledge* (London: Penguin, 1998).

11. Some computer scientists and cultural theorists may disagree with this characterization, but we have yet to find a compelling counterexample or counterargument.

12. This is already happening in some cases. For instance, a software program called ReporterMate received its first byline as the author of a news article in the *Guardian* in 2019. See "Political Donations Plunge to $16.7m," *The Guardian*, January 31, 2019, https://www.theguardian .com/australia-news/2019/feb/01/political-donations-plunge-to-167m-down-from-average -25m-a-year.

13. An excellent science-fictional exploration of AI authorship can be found in *Star Trek Voyager*, season 7, episode 20, "Author, Author," directed by David Livingston, written by Gene

Roddenberry, Rick Berman, and Michael Piller, in which a holographic author fights an exploitative publisher for the right to control a work of fiction he has produced.

14. This is a fascinating story in its own right, involving the celebrated author Oscar Wilde. For more information, see Burrow-Giles Lithographic Co. v. Sarony 111 U.S. 53 (1884).

15. As in chapter 2, we used the web-based GPT-2 app Write with Transformer to generate our text. At the time of writing, the site is freely available at https://transformer.huggingface.co /doc/gpt2-large.

16. Karen Hao, "Facebook's Ad-Serving Algorithm Discriminates by Gender and Race," *MIT Technology Review*, April 5, 2019, https://www.technologyreview.com/2019/04/05/1175 /facebook-algorithm-discriminates-ai-bias.

17. Shalini Kantayya, dir., *Coded Bias* (2020; Brooklyn, NY: 7th Empire Media); Ruha Benjamin, *Race after Technology: Abolitionist Tools for the New Jim Code* (Cambridge: Polity, 2019).

18. Yvonne D. Hodson and David Leibelshon, "Creating Databases with Students," *School Library Journal* 32, no. 1 (1985): 13.

19. Rob Price, "Microsoft Took Its New A.I. Chatbot Offline after It Started Spewing Racist Tweets," *Slate*, March 24, 2016, https://slate.com/business/2016/03/microsoft-s-new-ai -chatbot-tay-removed-from-twitter-due-to-racist-tweets.html.

20. Unfortunately, Microsoft seems to have forgotten the lessons it learned from Tay; in 2023, the company rolled out a GPT-enhanced chat interface for its search engine, Bing, which rapidly went off the rails, providing disinformative and even threatening responses to queries by beta users. See James Vincent, "Microsoft's Bing Is an Emotionally Manipulative Liar, and People Love It," *The Verge*, February 15, 2023, https://www.theverge.com/2023/2/15/23599072 /microsoft-ai-bing-personality-conversations-spy-employees-webcams.

21. Jane Chung, "Racism In, Racism Out: A Primer on Algorithmic Racism." *Public Citizen*, n.d., https://www.citizen.org/article/algorithmic-racism.

22. Joy Buolamwini and Timnit Gebru, "Gender Shades: Intersectional Accuracy Disparities in Commercial Gender Classification," *Proceedings of Machine Learning Research* 81 (2018): 77–91.

23. E. M. Bender, T. Gebru, A. McMillan-Major, M. Mitchell, V. Prabhakaran, M. Díaz, and B. Hutchinson, "On the Dangers of Stochastic Parrots: Can Language Models Be Too Big?" in *Proceedings of the 2021 ACM Conference on Fairness, Accountability, and Transparency* (New York: ACM, 2021), 610–623.

24. Tom Simonite, "What Really Happened When Google Ousted Timnit Gebru," *Wired*, June 8, 2021, https://www.wired.com/story/google-timnit-gebru-ai-what-really-happened.

25. Sheera Frenkel and Cecilia Kang, *An Ugly Truth: Inside Facebook's Battle for Domination* (New York: Harper, 2021).

26. Sadie Gurman, "AP: Across US, Police Officers Abuse Confidential Databases," Associated Press, September 28, 2016, https://apnews.com/article/699236946e3140659fff8a2362e16f43.

27. Gurman, "Police Officers Abuse Confidential Databases."

28. As a matter of principle, we will not provide direct links to any of these cyberstalking resources.

29. Federal Trade Commission, "FTC Bans SpyFone and CEO from Surveillance Business and Orders Company to Delete All Secretly Stolen Data," September 1, 2021, https://www.ftc.gov /news-events/press-releases/2021/09/ftc-bans-spyfone-and-ceo-from-surveillance-business.

CHAPTER 4

1. There was a wealth of dystopian futuristic narratives during this era as well, from the *Planet of the Apes* film series to Ridley Scott's *Blade Runner* to the genre-defining Japanese animated film *Akira*. While many of these narratives also featured artificial intelligence and autonomous machines, they portrayed such technologies as more disruptive and destructive than the utopian stories mentioned above.

2. Despite the fact that neither author was raised in such a family, these narratives still played an important role in shaping our own expectations about the future of society and technology as we came of age along with desktop computers and the World Wide Web in the 1980s and '90s.

3. LeVar Burton, dir., *Smart House* (Disney, 1999), https://www.disneyplus.com/movies /smart-house/51pux3uBRZyh.

4. Gordon Moore, the former CEO of Intel, predicted in 1975 that the processing capacity of computer chips would double every two years, contributing to a geometric growth in their power and function. In the half century since then, actual technological development has roughly followed this predicted trend.

5. "HAPIfork by Jacques Lépine," HAPI.com, archived December 30, 2019 at https://web .archive.org/web/20191230205416/https://www.hapilabs.com/product/hapifork.

6. Carrie Mihalcik, "Squeezed Dry: Juicero Shuts Down," *CNET*, September 1, 2017, https:// www.cnet.com/home/kitchen-and-household/juicero-shuts-down-offers-refunds.

7. Jack Morse, "These $110 'Smart Flip Flops' Are Incredibly Dumb," Mashable, March 28, 2017, https://mashable.com/article/smart-flip-flops-are-actually-dumb.

8. "Numi 2.0," Kohler, https://www.kohler.com/en/products/toilets/shop-toilets/numi-2-0 -one-piece-elongated-smart-toilet-dual-flush-30754-pa.

9. Norbert Wiener, *The Human Use of Human Beings: Cybernetics and Society* (Cambridge, MA: Riverside Press, 1950), 9.

10. As scholars such as Donella Meadows have pointed out, feedback loops and their consequences are endemic to all complex systems, including social systems that long predate the advent of computer networks. See Donella H. Meadows, *Thinking in Systems* (White River Junction, VT: Chelsea Green Publishing, 2015).

11. "Amnesia33: Identify and Mitigate the Risk from Vulnerabilities Lurking in Millions of IoT and IT Devices," white paper, Forescout, 2020, https://www.forescout.com/resources /amnesia33-identify-and-mitigate-the-risk-from-vulnerabilities-lurking-in-millions-of-iot -ot-and-it-devices.

12. There is a sad irony to the fact that the open-source and free software ethos, which was so essential to the innovation and excitement that surrounded early internet development, has contributed to enduring flaws in its long-term building blocks. We don't intend this analysis in any way to denigrate the work of the thousands of coders who contributed to these projects, or the communal and altruistic spirit of their contributions. And we acknowledge that privatized code comes with its own set of flaws and limitations, many of which are mentioned elsewhere in this book.

13. In fact, as we were writing this chapter, an even more widespread cybersecurity vulnerability called Log4j was brought to public attention.

14. Said Jawad Saidi, et al, "A Haystack Full of Needles: Scalable Detection of IoT Devices in the Wild," arXiv, September 30, 2020, https://doi.org/10.48550/arXiv.2009.01880.

15. Geoffrey A. Fowler, "Alexa Has Been Eavesdropping on You This Whole Time," *Washington Post*, May 6, 2019, https://www.washingtonpost.com/technology/2019/05/06/alexa -has-been-eavesdropping-you-this-whole-time. Legally, the End User License Agreements (EULAs) associated with these devices constitute a form of consent, but purchasers rarely read, let alone understand, these EULAs, and other people in the room may not have technically consented in this way.

16. Natasha Lomas, "Get Popcorn for IOS 13's Privacy Pop-Ups of Creepy Facebook Data Grabs," TechCrunch, September 16, 2019, https://tcrn.ch/300BGoO. Facebook's secret use of Bluetooth beacons was only revealed publicly because Apple disabled this functionality in its new version of iOS, causing Facebook to complain about putatively anticompetitive behavior.

17. Nicholas Gabriel and Aaron Shapiro, "Failed Hybrids: The Death and Life of Bluetooth Proximity Marketing," *Mobile Media and Communication* 9, no. 3 (December 2020), https:// doi.org/10.1177/2050157920975836.

18. In the original film, the WOPR computer simulates global thermonuclear war, ultimately concluding that in such a "strange game, the only winning move is not to play." See John Badham, dir., *WarGames* (United Artists, 1983), https://www.amc.com/movies/wargames—1043773.

19. The business model associated with these data brokers is frequently referred to as *surveillance capitalism*, a term coined by Harvard professor Shoshana Zuboff.

20. A cookie, in technological terms, is a well-established piece of technology that allows advertisers and publishers to track web visitors as they travel from site to site.

21. For instance, a gigabyte of storage cost $300,000 in 1981, but the price fell to ten cents by 2010, and is about one or two cents at the time of writing.

22. Jon Keegan and Alfred Ng, "There's a Multibillion-Dollar Market for Your Phone's Location Data," *The Markup*, September 30, 2021, https://themarkup.org/privacy/2021/09/30/theres-a-multibillion-dollar-market-for-your-phones-location-data.

23. Jon Keegan and Alfred Ng, "The Popular Family Safety App Life360 Is Selling Precise Location Data on Its Tens of Millions of Users," *The Markup*, December 6, 2021, https://themarkup.org/privacy/2021/12/06/the-popular-family-safety-app-life360-is-selling-precise-location-data-on-its-tens-of-millions-of-user.

24. "Near Delivers Actionable Insights on Consumer Behavior at a Global Scale," Near.com, accessed August 21, 2023, https://near.com/about.

25. Nick Statt, "Amazon Plans to Install Always-on Surveillance Cameras in Its Delivery Vehicles," *The Verge*, February 3, 2021, https://www.theverge.com/2021/2/3/22265031/amazon-netradyne-driveri-surveillance-cameras-delivery-monitor-packages.

26. Eric Rosenbaum, "IBM Artificial Intelligence Can Predict with 95% Accuracy Which Workers Are about to Quit Their Jobs," CNBC, April 3, 2019, https://www.cnbc.com/2019/04/03/ibm-ai-can-predict-with-95-percent-accuracy-which-employees-will-quit.html.

27. As we've discussed, most consumers don't ever read, let alone actively consent to the terms of service attached to their technology products, and there's legitimate debate about how intrusive surveillance can be in public spaces. These issues are at the forefront of ongoing debates and battles over data privacy in nearly every industrialized country around the world.

28. Brian Huseman to Senator Edward Markey, July 1, 2022, https://www.markey.senate.gov/imo/media/doc/amazon_response_to_senator_markey-july_13_2022.pdf.

29. Bennett Cyphers, "Data Broker Veraset Gave Bulk Device-Level GPS Data to DC Government," Electronic Frontier Foundation, November 10, 2021, https://www.eff.org/deeplinks/2021/11/data-broker-veraset-gave-bulk-device-level-gps-data-dc-government.

30. Rhett Jones, "Secret Service Paid to Get Americans' Location Data without a Warrant, Documents Show," Gizmodo, August 17, 2020, https://gizmodo.com/secret-service-bought-access-to-americans-location-data-1844752501.

31. Drew Harwell and Craig Timberg, "How America's Surveillance Networks Helped the FBI Catch the Capitol Mob," *Washington Post*, April 2, 2021, https://www.washingtonpost.com/technology/2021/04/02/capitol-siege-arrests-technology-fbi-privacy.

32. Lauren Smiley, "A Brutal Murder, a Wearable Witness, and an Unlikely Suspect," *Wired*, September 17, 2019, https://www.wired.com/story/telltale-heart-fitbit-murder.

33. Minyvonne Burke, "Amazon's Alexa May Have Witnessed Alleged Florida Murder, Authorities Say," NBC News, November 2, 2019, https://www.nbcnews.com/news/us-news/amazon-s-alexa-may-have-witnessed-alleged-florida-murder-authorities-n1075621.

34. Lindsay O'Donnell, "More Than Half of IoT Devices Vulnerable to Severe Attacks." Threatpost, March 11, 2020, https://threatpost.com/half-iot-devices-vulnerable-severe-attacks/153609.

35. CBS Baltimore, "Maryland Department of Health Confirms Ransomware Attack Caused Disruption In COVID-19 Data Last Month," CBS News, January 12, 2022, https://www.cbsnews.com/baltimore/news/maryland-department-of-health-confirms-ransomware-attack-caused-disruption-in-covid-19-data-last-month.

CHAPTER 5

1. Tribune Media Wire, "Heat Camera at Tourist Attraction Spots Woman's Breast Cancer," Fox 8, October 24, 2019, https://fox8.com/news/heat-camera-at-tourist-attraction-spots-womans-breast-cancer.

2. Tribune Media Wire, "Heat Camera."

3. Flavia Rotondi, "Italy's Art Museums Are Open Again, and Big Data Is Watching," *Bloomberg*, July 14, 2021, https://www.bloomberg.com/news/articles/2021-07-15/new-museum-technology-collects-data-on-viewing-habits.

4. It is unclear how the curators themselves feel about this initiative, or whether museumgoers will relish the idea of having their gazes tracked, even anonymously (if such a thing is truly possible), as they try to enjoy Italy's cultural heritage. There's also a question about whether these kinds of optimizations actually benefit culture at large—sometimes, the weird, unpopular artworks end up having the greatest impact, especially for marginalized viewers. Even in these best-case scenarios, biometrics can complicate ethical decision making.

5. Michael E. Ruane, "Faces of the Dead Emerge from Lost African American Graveyard." *Washington Post*, July 9, 2021, https://www.washingtonpost.com/history/2021/07/09/african-american-cemetery-catoctin-enslaved-faces.

6. Will Knight, "Clearview AI Has New Tools to Identify You in Photos," *Wired*, October 4, 2021, https://www.wired.com/story/clearview-ai-new-tools-identify-you-photos.

7. Juliette Pearse, "Clearview AI: Innovating Facial Recognition," *Time*, April 26, 2021, https://time.com/collection/time100-companies/5953748/clearview-ai/. On the day this sentence was written, four US senators wrote a letter to the secretary of homeland security urging the DHS to "stop use of facial recognition tools, including Clearview AI's products." Edward Markey, Pramila Jayapal, Jeffrey Merkley, and Ayanna Pressley to Alejandro Mayorkas, February 9, 2022, https://pressley.house.gov/sites/pressley.house.gov/files/Letters%20to%20Federal%20Agencies%20on%20Clearview%20AI.pdf. By the time you read this book, the debate will doubtless have evolved considerably.

8. Elizabeth Dwoskin, "Israel Escalates Surveillance of Palestinians with Facial Recognition Program in West Bank," *Washington Post*, November 5, 2021, https://www.washingtonpost.com/world/middle_east/israel-palestinians-surveillance-facial-recognition/2021/11/05/3787bf42-26b2-11ec-8739-5cb6aba30a30_story.html.

9. Thomas Lum and Michael A. Weber, "Human Rights in China and U.S. Policy: Issues for the 117th Congress," Congressional Research Service, March 31, 2021, https://crsreports.congress.gov/product/pdf/R/R46750.

10. Paul Mozur, "One Month, 500,000 Face Scans: How China Is Using A.I. to Profile a Minority," *New York Times*, April 14, 2019, https://www.nytimes.com/2019/04/14/technology/china-surveillance-artificial-intelligence-racial-profiling.html.

11. Harwell and Dou, "Huawei Tested AI Software."

12. Tom Phillips, "Moscow Metro to Roll Out Biometric Ticketing across Entire Network," *NFCW* (blog), September 13, 2021, https://www.nfcw.com/transit-ticketing-today/moscow-metro-to-roll-out-biometric-ticketing-across-entire-network.

13. "Global State of Democracy 2022: Forging Social Contracts in a Time of Discontent," Global State of Democracy Initiative, accessed August 5, 2022, https://idea.int/democracytracker/gsod-report-2022.

14. Kim Lyons, "The IRS Will Soon Make You Use Facial Recognition to Access Your Taxes Online," *The Verge*, January 20, 2022, https://www.theverge.com/2022/1/20/22893057/irs-facial-recognition-taxes-online-idme-identity.

15. Ken Klippenstein and Sara Sirota, "The Taliban Have Seized US Military Biometric Devices," *The Intercept*, August 17, 2021, https://theintercept.com/2021/08/17/afghanistan-taliban-military-biometrics.

16. This is one of the many ways in which we can understand the challenges of data and technology as analogous to those surrounding issues like climate change and public health.

17. Michael Kwet, "The Microsoft Police State: Mass Surveillance, Facial Recognition, and the Azure Cloud," *The Intercept*, July 14, 2020, https://theintercept.com/2020/07/14/microsoft-police-state-mass-surveillance-facial-recognition.

18. As this chapter was being written, Apple released iOS 15.4, a new operating system for its mobile devices that allows users to unlock them with facial recognition while wearing a mask.

19. Ryan Webster, Julien Rabin, Loic Simon, and Frederic Jurie, "This Person (Probably) Exists: Identity Membership Attacks against GAN Generated Faces," arXiv, June 13, 2021, https://doi.org/10.48550/arXiv.2107.06018.

20. Megan Cassidy, "San Francisco Police Linked a Woman to a Crime Using DNA from Her Rape Exam, D.A. Boudin Says," *San Francisco Chronicle*, February 14, 2022, https://www.sfchronicle.com/sf/article/San-Francisco-police-linked-a-woman-to-a-crime-16918673.php.

21. Homer, *The Odyssey*, translated by William Butler (London: A. C. Fifield, 1897), https://en.wikisource.org/wiki/The_Odyssey_(Butler); William Shakespeare, *The Merchant of Venice*, act 2, sc. 2.

22. Dinitia Smith and Nicholas Wade, "DNA Test Finds Evidence of Jefferson Child by Slave," *New York Times*, November 1, 1998, https://www.nytimes.com/1998/11/01/us/dna-test-finds-evidence-of-jefferson-child-by-slave.html.

23. Jacqueline Mroz, "When an Ancestry Search Reveals Fertility Fraud," *New York Times*, February 28, 2022, https://www.nytimes.com/2022/02/28/health/fertility-doctors-fraud-rochester.html.

24. Justin Hendrix, "Purported Deepfake of Ukrainian President Zelensky Aired on Television Station Website," Tech Policy Press, March 16, 2022, https://techpolicy.press/purported-deepfake-of-ukrainian-president-zelensky-aired-on-television-state-website.

25. While this technology is new, the premise itself is not. As the philosopher Hannah Arendt warned, "defactualization"—what is now more frequently referred to as "post-truth"—can be a potent weapon for authoritarian regimes. See Hannah Arendt, "Lying in Politics," in *Crises of the Republic* (New York: Harcourt Brace, 1972), 1–48.

26. Michael Zhang, "This AI Turns Pixel Faces into 'Photos,'" PetaPixel, June 20, 2020, https://petapixel.com/2020/06/20/this-ai-turns-pixel-faces-into-photos.

27. Chicken3gg (@Chicken3gg), Twitter, June 20, 2020, https://twitter.com/Chicken3gg/status/1274314622447820801.

28. Alistair Barr, "Google Mistakenly Tags Black People as 'Gorillas,' Showing Limits of Algorithms," *Wall Street Journal*, July 1, 2015, https://www.wsj.com/articles/BL-DGB-42522; "Twitter Investigates Racial Bias in Image Previews," BBC News, September 21, 2020, https://www.bbc.com/news/technology-54234822.

29. Sinduja Rangarajan, "Hey Siri—Why Don't You Understand More People Who Talk Like Me?" *Mother Jones* (blog), February 23, 2021, https://www.motherjones.com/media/2021/02/digital-assistants-accents-english-race-google-siri-alexa.

30. Haje Jan Kamps, "Sayso Is Launching an API to Dial Down People's Accents a Wee Bit," TechCrunch, March 14, 2022, https://techcrunch.com/2022/03/14/sayso-accent-changing/?tpcc=tcplustwitter&guccounter=1. While the technology itself is new, the promise (and peril) of accent erasure has been an important dimension of the immigrant experience in the United States and elsewhere for centuries, if not longer.

31. Sayso, accessed August 18, 2023, https://sayso.ai.

32. Salman Ahmed, Cameron T. Nutt, Nwamaka D. Eneanya, Peter P. Reese, Karthik Sivashanker, Michelle Morse, Thomas Sequist, et al., "Examining the Potential Impact of Race Multiplier Utilization in Estimated Glomerular Filtration Rate Calculation on African-American Care Outcomes," *Journal of General Internal Medicine* 36, no. 2 (February 2021): 464–471, https://doi.org/10.1007/s11606-020-06280-5.

33. Laure Wynants, "Prediction Models for Diagnosis and Prognosis of COVID-19: Systematic Review and Critical Appraisal," *British Medical Journal* 369, no. 8242 (2020), https://doi.org/10.1136/bmj.m1328.

34. Yilun Wang and Michal Kosinski, "Deep Neural Networks Are More Accurate Than Humans at Detecting Sexual Orientation from Facial Images," *Journal of Personality and Social Psychology* 114, no. 2 (2018): 246–257, https://doi.org/10.1037/pspa0000098.

35. "The Truth About Lie Detectors (aka Polygraph Tests)," American Psychological Association, August 5, 2004, https://www.apa.org/topics/cognitive-neuroscience/polygraph. In case this seems unclear, "highly problematic" is academic jargon for "this is a really bad idea with catastrophic consequences."

CHAPTER 6

1. The Strava Club, "The Stacey Abrams 50k," Strava, January 19, 2021, https://www.strava.com/clubs/231407/posts/14046003.

2. Sharonna Pearl, *About Faces: Physiognomy in Nineteenth-Century Britain* (Cambridge, MA: Harvard University Press, 2010).

3. *Quantified Self* (blog), https://quantifiedself.com/blog.

4. Gary Wolf, "The Quantified Self," filmed September 2010 in Cannes, TED talk, 5:10, https://www.youtube.com/watch?v=OrAo8oBBFIo.

5. Google Trends, accessed July 20, 2023, https://trends.google.com/trends/explore?date=all&geo=US&q=%22quantified%20self%22&hl=en.

6. IDC, "Wearable Devices Market Share," archived at https://web.archive.org/web/20220520201529/https://www.idc.com/promo/wearablevendor.

7. "Wearable Devices Are Connecting Health Care to Daily Life," *The Economist*, May 2, 2022, https://www.economist.com/technology-quarterly/2022/05/02/wearable-devices-are-connecting-health-care-to-daily-life.

8. Including, we're sorry to report, at least one institution of higher education. As of this writing, the Moodbeam home page listed the University of Hull as a client.

9. See, for example, Suzanne Bearne, "A Wristband That Tells Your Boss If You Are Unhappy," BBC News, January 18, 2021, https://www.bbc.com/news/business-55637328, and Jonathan Margolis, "The Wearable Device That Allows You to Log Your Mood," *Financial Times*, August 15, 2019, https://www.ft.com/content/e2829a66-be77-11e9-9381-78bab8a70848.

10. Feel Therapeutics, "Our Vision," accessed July 20, 2023, https://www.feeltherapeutics.com/about-feel.

11. See United Language Group, "The Sapir Whorf Hypothesis and Language's Effect on Cognition," *ULG's Language Solutions Blog*, April 2017, https://www.unitedlanguagegroup.com/blog/the-sapir-whorf-hypothesis-and-languages-effect-on-cognition.

12. Tim Culpan, "Beware That Nocebo Strapped to Your Wrist," *Bloomberg*, December 15, 2021, https://www.bloomberg.com/opinion/articles/2021-12-15/wrist-size-fitness-gadgets-make-for-great-gifts-but-beware-of-the-nocebo-effect.

13. Renee Engeln, Ryan Loach, Megan N. Imundo, and Anne Zola, "Compared to Facebook, Instagram Use Causes More Appearance Comparison and Lower Body Satisfaction in College Women," *Body Image* 34 (2020): 38–45, https://doi.org/10.1016/j.bodyim.2020.04.007.

14. Shannon Bond, "Lawmakers Push Facebook to Abandon Instagram for Kids, Citing Mental Health Concerns," National Public Radio, September 15, 2021, https://www.npr.org/2021/09/15/1037222495/lawmakers-push-facebook-to-abandon-instagram-for-kids-citing-mental-health-conce.

15. See "The Facebook Files," *Wall Street Journal*, accessed August 21, 2023, https://www.wsj .com/articles/the-facebook-files-11631713039?mod=article_inline.

16. Pratiti Raychoudhury, "What Our Research Really Says about Teen Well-Being and Instagram," Meta, September 26, 2021, https://about.fb.com/news/2021/09/research-teen -well-being-and-instagram.

17. Abby Ohlheiser, "TikTok Changes the Shape of Some People's Faces without Asking," *MIT Technology Review*, June 10, 2021, https://www.technologyreview.com/2021/06/10/1026074 /tiktok-mandatory-beauty-filter-bug.

18. Danielle Braff, "Plastic Surgeons Say Business Is Up, Partly Because Clients Don't Like How They Look on Zoom," *Washington Post*, December 7, 2020, https://www.washingtonpost .com/road-to-recovery/plastic-surgery-cosmetic-covid-zoom/2020/12/07/6283e6d2-35a2 -11eb-b59c-adb7153d10c2_story.html.

19. In an academic publication, we used the term *informatic subjectivity*, which might be a bit more clinically accurate, but it's less memorable, and less fun to say, than *algo-vision*. See Sinnreich and Gilbert, "Carrier Wave Principle."

20. Taylor Lorenz, "'Algospeak' Is Changing Our Language in Real Time," *Washington Post*, April 8, 2022, https://www.washingtonpost.com/technology/2022/04/08/algospeak -tiktok-le-dollar-bean.

21. Sherry Turkle, *The Second Self: Computers and the Human Spirit*, 20th anniversary edition (Cambridge, MA: MIT Press, 2005), 29.

22. See Chris Fullwood, Sally Quinn, Josephine Chen-Wilson, Darren Chadwick, and Katie Reynolds, "Put On a Smiley Face: Textspeak and Personality Perceptions," *Cyberpsychology, Behavior, and Social Networking* 18, no. 3 (2015): 147–151, and Juan María Tellería, "English and Leetspeak: A Step towards Global Nerdism," *Fòrum de Recerca* 17 (2012): 653–668.

23. danah boyd and Alice Marwick, "Social Steganography: Privacy in Networked Publics" (paper, International Communication Association conference, Boston, MA, May 28, 2011).

24. Nicole Amare and Alan Manning, "Writing for the Robot: How employer search tools have influenced résumé rhetoric and ethics," *Business Communication Quarterly* 72, no. 1 (2009): 35–60.

25. Forlano herself has written extensively on this subject. See Laura Forlano, "Data Rituals in Intimate Infrastructures: Crip Time and the Disabled Cyborg Body as an Epistemic Site of Feminist Science," *Catalyst: Feminism, Theory, Technoscience* 3, no. 2 (2017): 1–28, and Laura Forlano, "Posthuman Futures: Connecting/Disconnecting the Networked (Medical) Self," in *A Networked Self and Human Augmentics, Artificial Intelligence, Sentience*, ed. Zizi Papacharissi (Oxfordshire: Routledge, 2018), 39–50.

26. Renkai Ma and Yubo Kou, "'How Advertiser-Friendly Is My Video?': YouTuber's Socioeconomic Interactions with Algorithmic Content Moderation," *Proceedings of the ACM on Human-Computer Interaction* 5, no. CSCW2 (2021): 1–25.

27. Ireti Akinrinade and Joan Mukogosi, "Strategic Knowledge," *Points* (blog), *Data and Society*, July 14, 2021, https://points.datasociety.net/strategic-knowledge-6bbddb3f0259.

28. Zuxuan Wu, Ser-Nam Lim, Larry Davis, and Tom Goldstein, "Making an Invisibility Cloak: Real World Adversarial Attacks on Object Detectors," in *Computer Vision–ECCV 2020: 16th European Conference, Glasgow, UK, August 23–28, 2020, Proceedings, Part IV*, eds. Andrea Vedaldi, Horst Bischof, Thomas Brox, and Jan-Michael Frahm (Basel: Springer Nature, 2020), 1–17; Lauren Valenti, "Can Makeup Be an Anti-Surveillance Tool?" *Vogue*, June 12, 2020, https://www.vogue.com/article/anti-surveillance-makeup-cv-dazzle-protest; Melissa Locker, "To Thwart Face Recognition, Maybe Just Wear Juggalo Makeup," *Fast Company*, n.d, https://www.fastcompany.com/90373952/to-thwart-face-recognition-maybe-just -wear-juggalo-makeup.

29. Steve Hendrix, "Traffic-Weary Homeowners and Waze Are at War, Again. Guess Who's Winning?" *Washington Post*, June 5, 2016, https://www.washingtonpost.com/local/traffic -weary-homeowners-and-waze-are-at-war-again-guess-whos-winning/2016/06/05/c466df46 -299d-11e6-b989-4e5479715b54_story.html.

30. Jennifer Sandlin, "Home Workers Use Clever Tricks to Resist Productivity Monitoring Technology," Boing Boing, August 26, 2022, https://boingboing.net/2022/08/26/home -workers-use-clever-tricks-to-resist-productivity-monitoring-technology.html.

31. Aarian Marshall, "Gig Workers Gather Their Own Data to Check the Algorithm's Math," *Wired*, February 24, 2021, https://www.wired.com/story/gig-workers-gather-data-check -algorithm-math.

32. David Bauder, "AP Says It Will No Longer Name Suspects in Minor Crimes," Associated Press, June 15, 2021, https://apnews.com/article/crime-technology-df0a7cd66590 d9cb29ed1526ec03b58f.

33. Bobby Allyn, "Snapchat Ends 'Speed Filter' That Critics Say Encouraged Reckless Driving," NPR, June 17, 2021, https://www.npr.org/2021/06/17/1007385955/snapchat-ends-speed -filter-that-critics-say-encouraged-reckless-driving.

34. Adam Smith, "Amazon Stops Telling People What They Have Bought in Emails," *The Independent*, June 2, 2020, https://www.independent.co.uk/tech/amazon-order-email-confirma tion-shipping-details-a9543966.html.

35. "FTC Explores Rules Cracking Down on Commercial Surveillance and Lax Data Security Practices," Federal Trade Commission, August 11, 2022, https://www.ftc.gov /news-events/news/press-releases/2022/08/ftc-explores-rules-cracking-down-commercial -surveillance-lax-data-security-practices; "Dark Patterns in Social Media Platform Interfaces: How to Recognise and Avoid Them," European Data Protection Board, March 14, 2022, https://edpb.europa.eu/system/files/2022-03/edpb_03-2022_guidelines_on_dark_patterns _in_social_media_platform_interfaces_en.pdf.

36. Rohit Chopra, "Truth in Advertising Event on the FTC's Remedial Authority," prepared remarks to TINA.org, January 11, 2021, https://www.ftc.gov/system/files/documents /public_statements/1589068/20210413_remarks_of_commissioner_chopra_at_tina.pdf.

37. Ray Kurzweil, *The Singularity Is Near: When Humans Transcend Biology* (New York: Penguin, 2006).

CHAPTER 7

1. Jorge Luis Borges, *The Library of Babel* (Boston: David R. Godine; Enfield, UK: Airlift, 2000).

2. Jorge Luis Borges, *The Book of Sand* (New York: Dutton, 1977).

3. Jorge Luis Borges, *The Aleph* (Eastbourne: Gardners Books, 2000).

4. Jorge Luis Borges, "Of Exactitude in Science," in *Jorge Luis Borges: Collected Fictions*, trans. Andrew Hurley (New York: Penguin, 1998), 325.

5. Literally as we were writing this paragraph, privacy scholar Chris Gilliard tweeted a link to a BBC story about a technology project that aims to create a virtual doppelganger of the entire earth, with the comment "Borges would like a word." Gilliard (correctly) assumed that a significant portion of his 33,000 Twitter followers would understand exactly which Borges story he was referencing.

6. The term *mixed reality* is often used to describe a hybrid of virtual reality (a completely immersive, computer-generated experience) and augmented reality (the superimposition of digital information as a layer over physical reality). But there are no clear demarcations between these concepts, so we use the term *mixed reality* in its broadest sense to encompass the entire spectrum of computer-enhanced immersive sensory experiences.

7. Donna Haraway, "Situated Knowledges: The Science Question in Feminism and the Privilege of Partial Perspective," *Feminist Studies* 14, no. 3 (1988): 575–599, https://doi.org/10.2307/3178066.

8. For more spurious correlations, check out her blog: http://www.dontyoulovedata.com/data-explore/fun-correlations.html.

9. Kyshia Henderson, Samuel Powers, Michele Claibourn, and Sophie Trawalter, "Confederate Monuments and the History of Lynching in the American South: An Empirical Examination," *Proceedings of the National Academy of Sciences* 118, no. 42 (2021): e2103519118, https://www.pnas.org/doi/full/10.1073/pnas.2103519118.

10. The Desert Fireball Network project, Fireballs in the Sky, accessed August 26, 2023, https://fireballsinthesky.com.au/the-research.

11. Mikayla Mace Kelly, "UArizona-Led Team Finds Nearly 500 Ancient Ceremonial Sites in Southern Mexico," *news.arizona.edu*, October 25, 2021, https://news.arizona.edu/story/uarizona-led-team-finds-nearly-500-ancient-ceremonial-sites-southern-mexico.

12. Molly Porter, "NASA, Partners Develop 'Lunar Backpack' Technology to Aid New Moon Explorers," National Air and Space Administration, April 20, 2022, https://www.nasa.gov/centers/marshall/news/releases/2022/nasa-partners-develop-lunar-backpack-technology-to-aid-new-moon-explorers.html.

13. T. Lauvaux, C. Giron, M. Mazzolini, A. D'aspremont, R. Duren, D. Cusworth, D. Shindell, et al., "Global assessment of oil and gas methane ultra-emitters." *Science* 375, no. 6580 (2022): 557–561.

14. Vimal Patel, "Timber Poachers Set a Forest on Fire. Tree DNA Sent One to Prison," *New York Times*, November 10, 2021, https://www.nytimes.com/2021/11/10/us/justin-wilke-maple -fire-tree-dna-sentenced.html?referringSource=articleShare.

15. ShotSpotter, "ShotSpotter's Precision Policing Platform Contributes to Positive Outcomes," archived at https://web.archive.org/web/20220221155349/https://www.shotspotter.com; ShotSpotter rebranded as SoundThinking in April 2023, perhaps in response to the negative publicity associated with the *Motherboard* article.

16. Todd Feathers, "Police Are Telling ShotSpotter to Alter Evidence from Gunshot-Detecting AI," *Vice*, July 26, 2021, https://www.vice.com/en/article/qj8xbq/police-are-telling-shot spotter-to-alter-evidence-from-gunshot-detecting-ai.

17. "SoundThinking Responds to False Claims," SoundThinking, accessed July 20, 2023, https://www.soundthinking.com/soundthinking-responds-to-false-claims.

18. Amin Sarafraz and Brian K. Haus, "A Structured Light Method forUunderwater Surface Reconstruction," *ISPRS Journal of Photogrammetry and Remote Sensing* 114 (2016): 40–52, https://www.sciencedirect.com/science/article/abs/pii/S0924271616000290; Julian Isering-hausen, B. Goldlücke, N. Pesheva, S. Iliev, A. Wender, M. Fuchs, and M. B. Hullin, "4D Imaging through Spray-On Optics," *ACM Transactions on Graphics* 36, no. 4 (2017): 11; J. Xiong and W. Heidrich, "In-the-Wild Single Camera 3D Reconstruction through Moving Water Surfaces," https://openaccess.thecvf.com/content/ICCV2021/papers/Xiong_In-the -Wild_Single_Camera_3D_Reconstruction_Through_Moving_Water_Surfaces_ICCV _2021_paper.pdf.

19. Full disclosure: one such science-fictional book is a 2023 novel coauthored by one of the authors of this book. See R. A. Sinn, *A Second Chance for Yesterday* (Oxford: Solaris, 2023).

20. Valuates Reports, "Augmented and Virtual Reality (AR & VR) Market to Reach USD 454.73 Billion by 2030 with a CAGR of 40.7%," PR Newswire, June 22, 2022, https://www .prnewswire.com/news-releases/augmented-and-virtual-reality-ar--vr-market-to-reach-usd -454-73-billion-by-2030-with-a-cagr-of-40-7--valuates-reports-301573069.html.

21. "Meta" is a reference to *Snow Crash*, a groundbreaking 1992 dystopian novel by Neal Stephen-son about the potential dangers of MR. Cf. Neal Stephenson, *Snow Crash* (New York: Del Rey, 1992). Why Meta (formerly Facebook) would knowingly reshape its corporate brand around a dystopian vision of the future is an interesting question.

22. "Announcing Project Aria: A Research Project on the Future of Wearable AR," Meta, Sep-tember 16, 2020, https://about.fb.com/news/2020/09/announcing-project-aria-a-research -project-on-the-future-of-wearable-ar.

23. "Announcing Project Aria."

24. Sally A. Applin and Catherine Flick, "Facebook's Project Aria Indicates Problems for Respon-sible Innovation When Broadly Deploying AR and Other Pervasive Technology in the Commons," *Journal of Responsible Technology* 5 (May 2021): 100010, https://doi.org /10.1016/j.jrt.2021.100010.

25. We are well aware that private companies already surveil us physically and virtually throughout most of our daily activities, as we've discussed elsewhere in this book. But MR promises a qualitatively different scale of potential data collection, rooted in real-time, three-dimensional, large-scale sensor networks subject to immediate processing and reintegrated directly into the immersive experiences of billions of others. Think of this as the midpoint between a CCTV network and the Matrix.

26. See Lawrence Lessig, "Code and the Commons," keynote address, Conference on Media Convergence, Fordham University Law School, New York, February 9, 1999.

27. Jemima Kiss, "Google Admits Collecting Wi-Fi Data through Street View Cars," *The Guardian*, May 14, 2010, https://www.theguardian.com/technology/2010/may/15/google-admits-storing-private-data.

28. See Mahir Zaveri, "N.Y.P.D. Robot Dog's Run Is Cut Short After Fierce Backlash," *New York Times*, April 28, 2021, https://www.nytimes.com/2021/04/28/nyregion/nypd-robot-dog-backlash.html, and Yufeng Kok, "Autonomous Robots Check on Bad Behaviour in Singapore's Heartland," *Straits Times*, September 5, 2021, https://www.straitstimes.com/singapore/autonomous-robots-checking-on-bad-behaviour-in-the-heartland.

29. Zaveri, "N.Y.P.D. Robot Dog."

30. See Ed Pilkington, "NYPD Settles Lawsuit after Illegally Spying on Muslims," April 5, 2018, *The Guardian*, https://www.theguardian.com/world/2018/apr/05/nypd-muslim-surveillance-settlement; New York Civil Liberties Union, "A Closer Look at Stop-and-Frisk in NYC," 2023, https://www.nyclu.org/en/closer-look-stop-and-frisk-nyc; and Kade Crockford, "How Is Face Recognition Surveillance Technology Racist?," American Civil Liberties Union, June 16, 2020, https://www.aclu.org/news/privacy-technology/how-is-face-recognition-surveillance-technology-racist.

31. Emily Hopkins and Melissa Sanchez, "Chicago's 'Race-Neutral' Traffic Cameras Ticket Black and Latino Drivers the Most," ProPublica, January 11, 2022, https://www.propublica.org/article/chicagos-race-neutral-traffic-cameras-ticket-black-and-latino-drivers-the-most.

32. Cybersecurity experts may differ with us on this point and say that the difference between Easter eggs and malware is more substantive than semantic. They're probably right, but our discussion in this section applies just as well in either case.

33. Again, when we use this metaphor, we're not discussing only the specific form of malware that's referred to by cybersecurity experts as *trojans*, but rather a much more common and more pernicious set of vulnerabilities that pervades most computing hardware and software.

34. Ed Fries, "Microsoft Easter Egg Discussion with Ed Fries," interview by Jeremy Sachs, Rezmason, January 27, 2022, https://rezmason.github.io/excel_97_egg/research/ed_fries_1_27_22.html.

35. Microsoft's "culture of Easter eggs" effectively ended in 2002 with its "trustworthy computing" initiative, announced by CEO Bill Gates, which aimed to purge its code of vulnerabilities and flaws, including Easter eggs. Despite this initiative, Microsoft continued to ship

bloatware, products with embedded Easter eggs, and—as the Stuxnet story exemplifies—platforms that contained major security vulnerabilities.

36. Ellen Nakashima and Joby Warrick, "Stuxnet Was Work of U.S. and Israeli Experts, Officials Say," *Washington Post*, June 2, 2012, https://www.washingtonpost.com/world/national-security/stuxnet-was-work-of-us-and-israeli-experts-officials-say/2012/06/01/gJQAlnEy6U_story.html.

37. Andy Greenberg, "Biohackers Encoded Malware in a Strand of DNA," *Wired*, August 10, 2017, https://www.wired.com/story/malware-dna-hack.

38. As we've discussed throughout this book, all of these things are already happening to many people in their ostensibly offline lives, especially among marginalized and politically vulnerable communities; our point here is simply that these circumstances will become normative and ubiquitous for everyone.

39. Jonathan Zittrain, John Bowers, and Clare Stanton, "The Paper of Record Meets an Ephemeral Web: An Examination of Linkrot and Content Drift within the New York Times," SSRN, submitted April 27, 2021, http://dx.doi.org/10.2139/ssrn.3833133.

40. Elizabeth E. Joh, "Dobbs Online: Digital Rights as Abortion Rights," SSRN, submitted September 22, 2022.

CHAPTER 8

1. Lisa Deaderick, "Ready to Imagine a Future That Centers People of Color? Play 'AfroRithms' Game at UC San Diego with Its Creators," *San Diego Union-Tribune*, October 2, 2022, https://www.sandiegouniontribune.com/columnists/story/2022-10-02/ready-to-imagine-a-future-that-centers-people-of-color-play-afrorithms-game-at-uc-san-diego-with-creators.

2. Lonny J. Avi Brooks, Daniel Sutko, Aram Sinnreich, and Ryan Wallace, "Afro-futuretyping Generation Starships and New Earths 05015 CE," *ETC: A Review of General Semantics* 72, no. 4 (2015): 410–426, http://www.jstor.org/stable/44857469.

3. Alexis de Tocqueville and John Canfield Spencer, *The Republic of the United States of America and Its Political Institutions, Reviewed and Examined*, vol. 1–2 (New York: E. Walker, 1849), 4.

4. Tocqueville and Spencer, 176.

5. David Brin, *The Postman* (New York: Spectra, 1985).

6. Aram Sinnreich, Arul Chib, and Jesse Gilbert, "Modeling Information Equality: Social and Media Latency Effects on Information Diffusion," *International Journal of Communication* 2 (2008): 28, https://www.researchgate.net/publication/283829867_Modeling_information_equality_Social_and_media_latency_effects_on_information_diffusion. Salience refers to things we need to know the most, provided as accurately and quickly as possible. Consider, for instance, the importance of access to information about the availability of vaccines and personal protective gear during the COVID-19 pandemic.

7. Wei Lu and Douglas Blanks Hindman, "Does the Digital Divide Matter More? Comparing the Effects of New Media and Old Media Use on the Education-Based Knowledge Gap," *Mass Communication and Society* 14, no. 2 (2011): 216–235.

8. Lucas Chancel, Thomas Piketty, Emmanuel Saez, and Gabriel Zucman, "World Inequality Report 2022," World Inequality Lab, https://wir2022.wid.world/www-site/uploads/2021/12/WorldInequalityReport2022_Full_Report.pdf.

9. International IDEA, "Global State of Democracy Report 2022," Global State of Democracy Initiative, https://idea.int/democracytracker/gsod-report-2022.

10. Voting Rights Project, "Everything You Always Wanted to Know About Redistricting, but Were Afraid to Ask," American Civil Liberties Union, April 2001, https://www.aclu.org/sites/default/files/FilesPDFs/redistricting_manual.pdf.

11. Henry, Lord Brougham, "Historical Sketches of Statesmen who Flourished in the Time of George III" (Paris: Baudry's European Library, 1839), 55, https://www.google.com/books/edition/Historical_Sketches_of_Statesmen_who_Flo/U_Y7AQAAMAAJ.

12. See Richard Kerbaj, *The Secret History of the Five Eyes: The Untold Story of the Shadowy International Spy Network, through Its Targets, Traitors and Spies* (London: John Blake, 2022).

13. Joseph Goodman, Angela Murphy, Morgan Streetman, and Mark Sweet, "Carnivore: Will It Devour Your Privacy?" *Duke Law and Technology Review* 1 (2001), https://scholarship.law.duke.edu/cgi/viewcontent.cgi?article=1027&context=dltr.

14. James Risen and Laura Poitras, "N.S.A. Report Outlined Goals for More Power," *New York Times*, November 23, 2013, https://www.nytimes.com/2013/11/23/us/politics/nsa-report-outlined-goals-for-more-power.html.

15. Glenn Greenwald, "NSA Collected US Email Records in Bulk for More than Two Years under Obama," *The Guardian*, June 27, 2013, https://www.theguardian.com/world/2013/jun/27/nsa-data-mining-authorised-obama. We should add that Greenwald is a controversial figure who in recent years has downplayed Russia's role in interfering with US elections as "wildly exaggerated hysteria"; rejected the premise that the coup attempt at the Capitol on January 6th, 2021, was an insurrection; and—perhaps worst of all—blocked one of the authors on Twitter. See Natalie Amato, "Useful Idiots: Glenn Greenwald on Russiagate and Mainstream Mediea," *Rolling Stone*, January 17, 2020, https://www.rollingstone.com/politics/politics-news/glenn-greenwald-russiagate-taibbi-useful-idiots-podcast-939380; Gustaf Kilander, "Trump Claims 'There Was No Insurrection' in Video Blasting Jan 6 Report," *The Independent*, December 24, 2022, https://www.independent.co.uk/news/world/americas/us-politics/donald-trump-jan-6-report-b2251090.html.

16. To name just one example, Russia carried out extensive hacking, disinformation, and cyberespionage operations in Ukraine prior to its 2022 invasion. See Andy Greenberg, "Russia's New Cyberwarfare in Ukraine Is Fast, Dirty, and Relentless," *Wired*, November 19, 2022, https://www.wired.com/story/russia-ukraine-cyberattacks-mandiant.

17. Terri Moon Cronk, "New DOD Chief Digital Artificial Intelligence Office Launches," Department of Defense, February 4, 2022, https://www.defense.gov/News/News-Stories /Article/Article/2923986/new-dod-chief-digital-artificial-intelligence-office-launches.

18. This is hardly a novel observation; as early as 2012, the *Economist* ran an article asking whether a "digital cold war" was brewing, and in early 2013, the Dutch politician Neelie Kroes, then vice president of the European Commission responsible for the digital agenda, gave a speech at the European Parliament entitled "Stopping a Digital Cold War." These are the threads of a discourse that became far more widespread a few months later, in the wake of the Snowden leaks. See L. S., "A Digital Cold War?" *The Economist*, December 14, 2012, https://www .economist.com/babbage/2012/12/14/a-digital-cold-war; Neelie Kroes, "Stopping a Digital Cold War," European Commission, February 28, 2013, https://ec.europa.eu/commission /presscorner/detail/en/SPEECH_13_167.

19. In many military and regulatory contexts, the term *cyber* is now used as a standalone noun representing any and all issues related to data-centric security, diplomacy, and warfare.

20. Ben Child, "North Korea May Have Hacked Sony for Kim Jong-Un Baiting in *The Interview*," *The Guardian*, December 1, 2014, https://www.theguardian.com/film/2014/dec/01 /north-korea-sony-hacked-the-interview.

21. Just as the original, post–World War II Cold War was used to justify extrajudicial assassinations and, ultimately, the surveillance, harassment, imprisonment, and killing of political leaders in the United States and elsewhere, the geopolitics of the digital cold war are being used to justify contemporary tactics of oppression. Thus, there is continuity from twentieth-century history to the present—but in the current case data plays a central role as a catalyst, broadening the scale of surveillance and harassment and accelerating the erosion of democratic norms. What remains consistent is the fact that leading world powers treat the rule of law and human rights as obstacles that may be sidestepped in the interest of their ideological ambitions on the global stage.

22. Cate Cadell, "China Harvests Masses of Data on Western Targets, Documents Show," *Washington Post*, December 31, 2021, https://www.washingtonpost.com/national-security /china-harvests-masses-of-data-on-western-targets-documents-show/2021/12/31/3981ce9c -538e-11ec-8927-c396fa861a71_story.html.

23. Maggie Miller, "US, UK Agencies Warn Russian Hackers Using 'Brute Force' to Target Hundreds of Groups," *The Hill*, July 1, 2021, https://thehill.com/policy/cybersecurity /561138-us-uk-agencies-warn-russian-hackers-using-brute-force-to-target-hundreds.

24. Thomas Brewster, "The FBI Is Secretly Using a $2 Billion Travel Company as a Global Surveillance Tool," *Forbes*, July 16, 2020, https://www.forbes.com/sites/thomasbrewster /2020/07/16/the-fbi-is-secretly-using-a-2-billion-company-for-global-travel-surveillance --the-us-could-do-the-same-to-track-covid-19/?sh=4df9094757eb.

25. David Pegg, Paul Lewis, Michael Safi, and Nina Lakhani, "FT Editor among 180 Journalists Identified by Clients of Spyware Firm," *The Guardian*, July 20, 2021, https://www

.theguardian.com/world/2021/jul/18/ft-editor-roula-khalaf-among-180-journalists-targeted
-nso-spyware.

26. Tilman Rodenhauser, "How International Law Applies to the Use of Information and Communications Technologies by States," International Committee of the Red Cross, March 8, 2023, https://www.icrc.org/en/document/how-international-law-applies-to-use -information-and-communications-technologies-by-states.

27. Nori Katagiri, "Why International Law and Norms Do Little in Preventing Non-state Cyber Attacks," *Journal of Cybersecurity* 7, no. 1 (2021): tyab009, https://academic.oup .com/cybersecurity/article/7/1/tyab009/6168044.

28. Shira Rubin, "Israeli Police Accused of Using Pegasus Spyware on Domestic Opponents of Netanyahu," *Washington Post*, January 18, 2022, https://www.washingtonpost.com/world /2022/01/18/israel-pegasus-activists-spyware.

29. Corin Faife, "Feds Are Tracking Phone Locations with Data Bought from Brokers," *The Verge*, July 18, 2022, https://www.theverge.com/2022/7/18/23268592/feds-buying-location -data-brokers-aclu-foia-dhs.

30. Daniel Boffey, "EU Border 'Lie Detector' System Criticized as Pseudoscience," *The Guardian*, November 2, 2018, https://www.theguardian.com/world/2018/nov/02/eu-border-lie -detection-system-criticised-as-pseudoscience.

31. Vera Bergengruen, "'We Became Like a Big Startup.' How Kyiv Adapted the City's Tech to Save Lives," *Time*, April 4, 2022, https://time.com/6163708/kyiv-digital-technology-app.

32. Joseph Cox, "Military Unit That Conducts Drone Strikes Bought Location Data from Ordinary Apps," *Vice*, March 4, 2021, https://www.vice.com/en/article/y3g97x/location-data -apps-drone-strikes-iowa-national-guard.

33. Byron Tau, "App Taps Unwitting Users Abroad to Gather Open-Source Intelligence," *Wall Street Journal*, June 24, 2021, https://www.wsj.com/articles/app-taps-unwitting-users -abroad-to-gather-open-source-intelligence-11624544026.

34. Lucas Ropek, "ICE Is Reportedly Using OnStar Location Data to Track Suspects," Gizmodo, April 1, 2021, https://gizmodo.com/ice-is-reportedly-using-onstar-location-data-to-track -s-1846598616.

35. Joseph Cox, "CDC Tracked Millions of Phones to See If Americans Followed COVID Lockdown Orders," *Vice*, May 3, 2022, https://www.vice.com/en/article/m7vymn/cdc -tracked-phones-location-data-curfews.

36. Rachel Pannett, "German Police Used a Tracing App to Scout Crime Witnesses. Some Fear That's Fuel for Covid Conspiracists," *Washington Post*, January 13, 2022, https://www .washingtonpost.com/world/2022/01/13/german-covid-contact-tracing-app-luca.

37. Cody Venzke, "School Surveillance and COVID-19," Center for Democracy and Technology, January 22, 2021, https://cdt.org/insights/school-surveillance-and-covid-19; Mark Keierleber, "'Really Alarming': The Rise of Smart Cameras Used to Catch Maskless Students

in US Schools," *The Guardian*, March 30, 2022, https://www.theguardian.com/world/2022/mar/30/smart-cameras-us-schools-artificial-intelligence.

38. John Stuart Mill, *On Liberty* (London: John W. Parker and Son, 1859).

39. Emma L. Briant, "Researching Influence Operations: 'Dark Arts' Mercenaries and the Digital Influence Industry," in *Oxford Handbook of Digital Diplomacy*, eds. Corneliu Bjola and Ilan Manor (Oxford: Oxford University Press, forthcoming).

40. One of the company's founders, Steve Bannon, was later Donald Trump's campaign manager, and is widely seen as a key architect of right-wing political movements around the world in the 2010s and '20s. Former Cambridge Analytica employee Christopher Wylie told CNN in 2018 that the company amounted to "Bannon's arsenal of weaponry to wage a culture war on America using military strategies." Curt Devine, Donie O'Sullivan, and Drew Griffin, "How Steve Bannon Used Cambridge Analytica to Further His Alt-Right Vision for America," CNN, updated May 16, 2018, https://www.cnn.com/2018/03/30/politics/bannon-cambridge-analytica/index.html.

41. We mention the OCEAN model in this chapter not as an endorsement of it, or any other reductive means of analyzing personality traits based on similar methods of analysis. As with all efforts to reduce the human experience to a set of data points, we suspect the method has limited utility overall. However, it's a matter of historical fact that CA used this model, and regardless of its broader accuracy and generalizability, it appears to have served its purpose in this particular case—namely, identifying individuals who are statistically more likely than the average person to be motivated by fear-based messaging.

42. L. M., "Justice Vindicated; or, an explanation of an Act of Parliament entitl'd, An Act for granting an Aid to his Majesty, by a Land Tax, &c." (London: T. Payne, 1725).

43. It can also be understood as a stark rejection of religious dogma premised on original sin, and therefore an ontological cornerstone for the development of secular society.

44. Yes, we are aware that these sound like made up names from a low-budget *Minority Report* knockoff movie. We assure you, they're completely real.

45. Isabelle Qian, Muyi Xiao, Paul Mozur, and Alexander Cardia, "Investigation into China's Expanding Surveillance State," *New York Times*, June 21, 2022, https://www.nytimes.com/2022/06/21/world/asia/china-surveillance-investigation.html.

46. Dhruv Mehrotra, Surya Mattu, Annie Gilbertson, and Aaron Sankin, "How We Determined Predictive Policing Software Disproportionately Targeted Low-Income, Black, and Latino Neighborhoods," Gizmodo, December 2, 2021, https://gizmodo.com/how-we-determined-predictive-policing-software-dispropo-1848139456.

47. Public analyses of the SCS, at least in English, tend to have a strong political bent, either overstating the extent of China's data regime to support a narrative of human rights abuses, or downplaying the scope to make it seem mundane and haphazard. Furthermore, though the system was initially planned for completion and nationwide mandatory enrollment by 2020, it is reportedly still a work in progress at the time of writing in 2022.

48. Katja Drinhausen and Vincent Brussee, "China's Social Credit System in 2021: From Fragmentation towards Integration," Merics, March 3, 2021, https://merics.org/en/report /chinas-social-credit-system-2021-fragmentation-towards-integration; Charlie Campbell, "How China Is Using Big Data to Create a Social Credit Score," *Time*, 2019, https://time.com /collection/davos-2019/5502592/china-social-credit-score.

49. Karen L. X. Wong and Amy Shields Dobson, "We're Just Data: Exploring China's Social Credit System in Relation to Digital Platform Ratings Cultures in Westernized Democracies," *Global Media and China* 4, no. 2, 2019, 220–232, https://doi.org/10.1177/2059436419856090.

50. Chris Riotta, "The Chairmen of Two House Committees Want Federal Agencies to Explain What They Are Doing with 'Billions of Data Points on Hundreds of Millions of Americans' Acquired from Data Brokers," FCW, August 17, 2022, https://fcw.com/congress/2022/08 /what-does-federal-government-buy-data-brokers/375963.

51. Chauncey Crail, "VPN Statistics and Trends in 2023," Forbes Advisor, updated February 9, 2023, https://www.forbes.com/advisor/business/vpn-statistics.

52. Adam Rawnsley, "These Nerds Saw Ukraine Invasion Start on Google Maps Before Vladimir Putin Said a Word," *Daily Beast*, https://www.thedailybeast.com/these-nerds-saw-ukraine -invasion-start-on-google-maps-before-vladimir-putin-said-a-word.

CONCLUSION

1. Jeremy B., Merrill and Will Oremus. "Five Points for Anger, One for a 'Like': How Facebook's Formula Fostered Rage and Misinformation," *Washington Post*, October 26, 2021, https:// www.washingtonpost.com/technology/2021/10/26/facebook-angry-emoji-algorithm.

2. In more precise academic terms, we would call this the *destabilization of epistemology* (with *epistemology* referring to how we know what we know), but it's a bit late in the game for us to start throwing around weighty philosophical language here.

3. Christine Clark, "A New Replication Crisis: Research That Is Less Likely to Be True Is Cited More," UC San Diego Today, May 21, 2021, https://today.ucsd.edu/story/a-new -replication-crisis-research-that-is-less-likely-be-true-is-cited-more.

4. Gabriel Grill, "Constructing certainty in machine learning: On the performativity of testing and its hold on the future," OSF Preprints, September 7, 2022, https://doi.org/10.31219 /osf.io/zekqv.

5. Office of Science and Technology Policy, "Blueprint for an AI Bill of Rights," White House, accessed July 20, 2023, https://www.whitehouse.gov/ostp/ai-bill-of-rights.

6. "President Biden Signs Executive Order to Strengthen Racial Equity and Support for Underserved Communities across the Federal Government," White House press release, February 16, 2023, https://www.whitehouse.gov/briefing-room/statements-releases/2023/02/16 /fact-sheet-president-biden-signs-executive-order-to-strengthen-racial-equity-and-support -for-underserved-communities-across-the-federal-government.

7. Eleanor Saitta (@dymaxion), "Not only must data sovereignty trump open data, but we need active pro-social countermeasures- a data justice movement: http://bit.ly/MUYkQi," Tweet, June 27, 2012, https://twitter.com/dymaxion/status/218062501999427586.

8. Lina Dencik, "Exploring Data Justice: Conceptions, Applications and Directions," *Information, Communication and Society* 22, no. 7: 873–881, https://doi.org/10.1080/1369118X.2019.1606268.

9. Tahu Kukutai and John Taylor, eds., *Indigenous Data Sovereignty Toward an Agenda*. (Canberra: ANU Press, 2016).

10. Amanda Anne Geppert and Laura Ellen Forlano, "Design for Equivalence: Agonism for Collective Emancipation in Participatory Design," *Proceedings of the Participatory Design Conference 2022*, vol. 1 (New York: Association for Computing Machinery, 2022), 158–168, https://doi.org/10.1145/3536169.3537790.

11. Any honest computer scientist will tell you this; when they build and publish AI and ML models, their papers are full of terms that emphasize "confidence intervals" and other markers of probabilistic inference. But when these models are marketed to the public at large, or discussed in journalistic coverage, those nuances tend to be obscured, and a more absolutist vision of algorithmic certainty prevails.

12. Deb Raji, "It's Time to Develop the Tools We Need to Hold Algorithms Accountable," Mozilla, February 2, 2022, https://foundation.mozilla.org/en/blog/its-time-to-develop-the-tools-we-need-to-hold-algorithms-accountable.

13. Reva Schwartz, Apostol Vassilev, Kristen Greene, Lori Perine, Andrew Burt, and Patrick Hall, "Towards a Standard for Identifying and Managing Bias in Artificial Intelligence," NIST Special Publication 1270, March 2022, https://doi.org/10.6028/NIST.SP.1270.

14. Margaret Mitchell, Simone Wu, Andrew Zaldivar, Parker Barnes, Lucy Vasserman, Ben Hutchinson, Elena Spritzer, et al, "Model Cards for Model Reporting," in *Proceedings of the Conference on Fairness, Accountability, and Transparency* (New York: Association for Computing Machinery, 2019), 220–229, https://doi.org/10.1145/3287560.3287596.

15. Including the lead author Margaret Mitchell and the AI ethicist Timnit Gebru, whose departure from Google we discussed in chapter 3. Another author of the paper was Deb Raji, whose work we discussed above.

16. "Bringing Cultural Context to Artificial Intelligence," Indigenous Voices of Wisdom (IVOW), n.d, https://www.ivow.ai/uploads/1/0/5/3/105390607/whitepaper_ikg_ivow_2.pdf.

17. Ben Beaumont-Thomas, "DJ /Rupture: How to Sing like a Sufi," *The Guardian*, March 26, 2013, https://www.theguardian.com/music/2013/mar/26/jace-clayton-dj-rapture.

18. Davar Ardalan, "Whose Culture Does Artificial Intelligence Represent and Which Human Needs Does It Prioritize?" Medium, May 31, 2022, https://idavar.medium.com/whose-culture-does-artificial-intelligence-represent-and-which-human-needs-does-it-prioritize-8a70ac26fb68.

19. Ardalan, "Whose Culture Does Artificial Intelligence Represent."

20. Hannah Sampson, "Coming to a Giant Airport Screen: Your Personal Flight Information," *Washington Post*, July 25, 2022, https://www.washingtonpost.com/travel/2022/07/25 /delta-tech-flight-info-screen.

21. Florimond Houssiau, Piotr Sapieżyński, Laura Radaelli, Erez Shmueli, and Yves-Alexandre de Montjoye, "Detrimental Network Effects in Privacy: A Graph-Theoretic Model for Node-Based Intrusions," *Patterns* 4, no. 1 (2023): 100662, https://doi.org/10.1016/j.patter .2022.100662.

22. Alexandra S. Levine, "Suicide Hotline Shares Data with For-Profit Spinoff, Raising Ethical Questions," Politico, January 28, 2022, https://www.politico.com/news/2022/01/28 /suicide-hotline-silicon-valley-privacy-debates-00002617.

23. Alexander W. Butler, Arthur Herman, and Daley Pagano, "Decrypting Crypto: Cryptocurrencies and the Quantum Computer Threat," Hudson Institute, April 2022, https://www .hudson.org/technology/decrypting-crypto-cryptocurrencies-and-the-quantum-computer -threat.

24. Joseph R. Biden, Jr., "National Security Memorandum on Promoting United States Leadership in Quantum Computing While Mitigating Risks to Vulnerable Cryptographic Systems," White House statement, May 4, 2022, https://www.whitehouse.gov/briefing-room /statements-releases/2022/05/04/national-security-memorandum-on-promoting-united -states-leadership-in-quantum-computing-while-mitigating-risks-to-vulnerable-crypto graphic-systems.

25. Biden, "National Security Memorandum."

26. Natasha Tiku, "The Google Engineer Who Thinks the Company's AI Has Come to Life," *Washington Post*, June 11, 2022, https://www.washingtonpost.com/technology/2022/06/11 /google-ai-lamda-blake-lemoine.

27. These debates stretch back to the dawn of modern computing, when legendary mathematician Alan Turing proposed his famous imitation game, arguing that for all intents and purposes, AI sentience and human sentience are functionally identical if they're indistinguishable to a human observer.

28. Émile P. Torres, "How AI Could Accidentally Extinguish Humankind," *Washington Post*, August 31, 2022, https://www.washingtonpost.com/opinions/2022/08/31/artificial -intelligence-worst-case-scenario-extinction.

29. Michael K. Cohen Marcus Hutter, and Michael A. Osborne, "Advanced Artificial Agents Intervene in the Provision of Reward," *AI Magazine* 43 (2022), 282–293, https://doi .org/10.1002/aaai.12064.

30. Max Hodak (@maxhodak_) "Humans are objectively bad with socialism (and on the contrary, capitalism is amazingly effective at advancing humanity), but machines might end up

reasoning about their identities and communities super differently. We are going to get so wrecked." Twitter, September 30, 2021, https://twitter.com/maxhodak_/status/144367316 3720806401?lang=en.

31. Gary Smith, "The AI Illusion—State-of-the-Art Chatbots Aren't What They Seem," *Mind-Matters*, March 21, 2022, https://mindmatters.ai/2022/03/the-ai-illusion-state-of-the-art -chatbots-arent-what-they-seem.

32. Ray Kurzweil, *The Singularity Is Near: When Humans Transcend Biology* (New York: Penguin, 2006).

Index

Chips, computer, 86
Chopra, Rohit, 157
Clarke, Arthur C., 41
Clayton, Jace, 228
Clearview AI, 115
Clifford, Maggie, 123–125
Clinton, Hillary, 15, 127
Cobain, Kurt, 65–66
Coded Bias, 70
Cold War, 190
Collective refusal, 75
Colonization, data, 172–173
Colorado, 221
Commons, privatized, 173
Como, Perry, 164–165
Computational flattening algorithms, 33
Computer Fraud and Abuse Act, 61, 95
Computers, personal, 25, 63, 70
Computer vision algorithms, 37, 169–170
Computer vision dazzle, 154
Confederate memorials, 164
Confirmation bias, 73, 218
Consent, user, 231–232
Content Authenticity Initiative, 22
Content drift, 181
Content moderation, 153
Cookie walls, 231
Copyrights, 64, 66–67
Corporate espionage, 50–51
Cosmetic surgery, 146
COVID-19 pandemic, 122, 131, 146, 198
Creative Commons, 173
Crime suspects, naming, 156
Criminal activities, 60, 106–107
Crisis Text Line, 231
Cross-referenced data, 50, 58–59
Crowdsourced stewardship, 166, 169
Crowley, Dennis, 100–101, 221
Cultural artifacts, data as, 216
Culture, quantization of, 217, 220
Curtin University, 165
Cybernetics, 90

Cyberphysical systems, 89–90, 93, 96–97, 99, 107–108
Cybersecurity, 95, 98–99, 106, 182–183, 191
Cyberspace, 100
Cyberstalking, 77–80
Cyberwarfare, 179, 192–194
CycleGAN, 127

Dark patterns, 157
Darwin, Charles, 25
Data. *See also* Biometric data; Data collection; Metadata
  abuse of access to, 77–78, 95–96
  affirmative rights, 222–223, 229
  as algorithms, 177
  ambient, 76
  and bias, 28–30
  big data, 49–51
  brokered, 99–103, 105–106, 122, 194, 197, 208
  as byproduct, 232
  characteristics of, 232–233
  cross-referenced, 50, 58–59
  as cultural artifacts, 216
  data analytics, 20
  data colonization, 172–173
  data journalism, 48
  data justice, 223–224
  data sovereignty, 224–225, 228
  data storage, 101
  data voids, 51
  defined, xiii
  ecological, 166
  emotional, 200
  and environment, 163–164, 167–172
  genetic, 123
  government, 5
  health, 59, 110–111, 131
  limitations of, 56
  malleability of, 197
  missing, 50–51, 182
  and objectivity, 25